T0136419

# Analog IC Design Techniques for Nanopower Biomedical Signal Processing

# RIVER PUBLISHERS SERIES IN BIOMEDICAL ENGINEERING

Volume 1

*Series Editor*

**Dinesh Kant Kumar**
*RMIT*
*Australia*

The "River Publishers Series in Biomedical Engineering" is a series of comprehensive academic and professional books which focus on the engineering and mathematics in medicine and biology. The series presents innovative experimental science and technological development in the biomedical field as well as clinical application of new developments.

Books published in the series include research monographs, edited volumes, handbooks and textbooks. The books provide professionals, researchers, educators, and advanced students in the field with an invaluable insight into the latest research and developments.

Topics covered in the series include, but are by no means restricted to the following:

- Biomedical engineering
- Biomedical physics and applied biophysics
- Bio-informatics
- Bio-metrics
- Bio-signals
- Medical Imaging

For a list of other books in this series, visit www.riverpublishers.com

# Analog IC Design Techniques for Nanopower Biomedical Signal Processing

**Chutham Sawigun**

Mahanakorn University of Technology
Thailand

**Wouter A. Serdijn**

Delft University of Technology
The Netherlands

River Publishers

*Published, sold and distributed by:*
River Publishers
Niels Jernes Vej 10
9220 Aalborg Ø
Denmark

River Publishers
Lange Geer 44
2611 PW Delft
The Netherlands

Tel.: +45369953197
www.riverpublishers.com

ISBN: 978-87-93379-29-9 (Hardback)
       978-87-93379-28-2 (Ebook)

©2016 River Publishers

# Contents

# Preface

This book deals with the design of analog integrated circuits that serve biomedical signal processing applications. It has resulted from our research works performed at Delft university of Technology. This book targets not only researchers in universities, but also professional designers of integrated circuit who deal with area and power consumption minimization of biomedical analog integrated circuits.

As the requirements for low power consumption and very small physical dimensions in portable, wearable and implantable medical devices are calling for integrated circuit design techniques using MOSFETs operating in the sub-threshold regime, this book first revisits some well-known circuit techniques that use CMOS devices biased in subthreshold in order to establish nano-power integrated circuit designs.

Based on the above findings, this book shows the development of a class-AB current-mode sample-and-hold circuit with an order of magnitude improvement in its figure of merit compared to other state-of-the-art designs. Also, the concepts and design procedures of 1) single-branch filters 2) follower-integrator-based lowpass filters and 3) modular transconductance reduction techniques for very low frequency filters are presented. Finally, to serve the requirement of a very large signal swing in an energy-based action potential detector, a nanopower class-AB current-mode analog multiplier is designed to handle input current amplitudes of more than 10 times the bias current of the multiplier circuit. The invented filter circuits have been fabricated in a standard 0.18 μm CMOS process in order to verify our circuit concepts and design procedures. Their experimental results are reported.

As the inventions found in this book encompass several types of both linear and non-linear circuits (viz. sample-and-hold circuits, filters and analog multipliers) that are both very compact and power efficient, we hope that this book can be beneficial to readers who are interested in subthreshold MOS circuit design for low-frequency, low-power biomedical applications.

# List of Figures

# List of Tables

# 1

---

# Introduction

---

## 1.1 Motivations

This book describes the IC design techniques for nanopower analog signal processing in biomedical applications presented in [1–9]. Aiming to maximize the capability of a single MOS transistor to serve the requirements of extremely low-power operation and reasonable physical size for wearable and implantable medical devices, both discrete and continuous-time design techniques are investigated and developed. The book also employs the nonlinear behavior of conventional circuit building blocks to serve stringent requirements such as the very low cutoff frequency needed for slowly varying bio-potentials filtering, or the very large signal swing demanded for action potential detection. These research activities are driven by the health-related motivations described in the following two subsections.

### 1.1.1 Understanding the Human Nervous System

Counting from the treatment of headache and gout for the Roman Emperor Claudius by using torpedo fish in AD46 [10], it has become almost two millennia since electricity has been used in medicine for the first time. The early treatment of neural disorders by means of electricity started without sufficient understanding of electricity and the human nervous system. One step forward was made after the discovery of the Leyden Jar by Van Musschenbroeck in 1746. This was evident from the observation that electric shocks cause muscle contractions by Franklin in 1774 (later confirmed again by Galvani in 1780) and the breakthough of electrotherapy by Cavallo in 1777 who used electricity to treat various diseases i.e., epilepsy, paralysis, deafness and blindness [11]. After having shed some light on this new way of medical treatment, electrical and electronic engineering developed a lot for almost a couple of centuries as can be seen from the existence of the invention of the transistor and wireless technology. Unfortunately, however, electrotherapy did not progress as rapidly as the progress in electrical and electronic engineering. The main obstacle that hindered its progress was a too limited understanding of the workings of the central nervous system.

After the Second World War, advances in semiconductor technology, microelectronics and micro-electro-mechanical systems (MEMS) allowed the use of electrical recording to study nervous systems at both the intra-cellular [12] and extra-cellular [13] levels by using microelectrodes [14]. Significant progress in understanding the nervous system has been made from these electronic interfaces and the knowledge obtained has been applied in several medical prosthetic devices such as retinal implants [15], cochlear implants (CIs) [16], deep brain stimulators (DBS) [17] and functional electrical stimulators (FES) [18].

Nowadays, it has become clear both in medicine and engineering that we have learned enough about the human nervous system to know that there is still much more to learn. In order to perform treatments effectively and minimize side effects, medical devices that allow simultaneous neural stimulation and recording are required. Neural responses are expected to be observed and studied while the body is being stimulated, and system parameters can then be adjusted to give appropriate stimuli [19]. To enable this mechanism, that is unfortunately not yet available in any current technology, smart implantable neural microsystems that involve microelectrodes and ultra low-power integrated circuits and systems need to be developed [12].

In addition, the technology development mentioned above is required in modern healthcare to alleviate the cost of future medical treatments as will be described in the next subsection.

## 1.1.2 Paradigm Shift in Bio-Medicine

Advancements in medical technology is one reason why the world's aging population is expanding rapidly. It has been predicted that the number of elderly people over 65 will reach 1 billion globally in 2030 [20]. Maintaining or improving quality of life for a longer lifespan is indeed very costly. Without careful prevention and management, healthcare expenses against age-related chronic diseases will become a financial burden for the global society. As indicated in the 2005 WHO Global Report, $558, $237 and $303 billion from national incomes (over the period of 2005–2015) of China, India and the Russian Federation, respectively, will be spent on the treatment of heart diseases, stroke and diabetes [21].

As a consequence, the idea of using information and communication technology to mobilize bio-medicine to address this problem has been developed. Traditional medicine focuses on diagnosis and treatments that are centralized at a hospital. An individual-centered healthcare system that focuses on the prevention of illness by early prediction of diseases and maintaining health on a daily basis could reduce healthcare costs in the future [22]. To enable this type of healthcare, body-area networks (BAN) (also called a body-sensor networks (BSN)) that require efficient, low-cost wearable and implantable medical technologies and devices need to be implemented [21].

The key to success of this new paradigm is to keep the operation and implementation costs of the aforementioned systems much lower than those of traditional medicine. These wearable and implantable devices need to be built from the cheapest technology and consume extremely low amounts of power. This demands for research activities in the same area as mentioned in Section 1.1.1 as well as smart sensor systems and technology, energy harvesting and ultra low-power signal processing.

Driven by the motivations mentioned above, this book deals with the design of signal processing analog circuits in complementary metal-oxide-semiconductor (CMOS) integrated circuit (IC) technology that serve biomedical electronic systems. Applications of the design techniques presented in this book can be seen from: 1) signal filtering in an analog CI processor, 2) signal filtering in an ECG detector or other neural recorders and 3) signal multiplication in an energy-based action potential detector.

## 1.2 Analog Signal Processing in Wearable and Implantable Devices

Apart from well-known vital signs (temperature, heart rate, respiratory rate, blood pressure), more complex biomedical (bio-potential) signals, i.e., electrocardiograms (ECGs), electrocorticograms (ECoGs), electroneurograms (ENGs), and electromyograms (EMGs), can be monitored using a typical ExG recording system as shown in Figure 1.1 [23]. All of the biomedical signals mentioned manifest themselves within the frequency range of approximately DC to 10 kHz. The amplitude of signals obtained from the recording electrode can vary in the range of a few $\mu$V to mV depending on the recording site and type of electrode used. These signals will be fed to an amplifier typically with a voltage gain of around 40 dB or more [23] before relaying the greater signal amplitudes to the next stage.

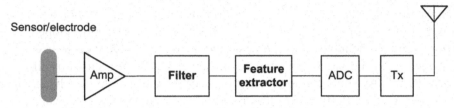

**Figure 1.1**    Typical neural recording architecture.

After signal amplification, the signal processing sections, i.e., filter and feature extractor are responsible for discrimination between the required signal and interfering signals and extracting relevant features of the signal, respectively. The extracted feature will be converted into a digital format by an analog-to-digital converter (ADC) before the transmission by the RF transmitter (Tx). There are also cases (found in wearable and portable devices) that after filtering the signal will be fed to the ADC directly. In multi-channel schemes, extracting only relevant features can help relax the resolution thereby lowering power consumption of the ADC. Feature extraction therefore leads to a reduced data rate and lower average transmission power [24].

This book describes design techniques to realize analog signal processors using standard CMOS integrated circuit technology in extremely compact and low-power fashions. It thus investigates the possibility to:

- **At the device level**: employ a single MOS device to perform signal processing functions in both the continuous and discrete-time domains.
- **At the circuit level**: apply class-AB techniques to enhance the dynamic range of subthreshold CMOS circuits further.

## 1.3 Nanopower CMOS IC Design Challenges

In CMOS technologies, nMOS and pMOS devices are used in combination with other passive devices (i.e., resistors, capacitors and inductors) to build the required signal processing functions. In this nanopower design context, MOS devices in a circuit will be operated in the subthreshold or weak inversion region at very low bias currents (in the range of a few nAs down to a few tens of pAs). This unavoidable constraint on the design leads to the following problems:

1. The voltage-to-current characteristic of the transistors is governed by an exponential function which is highly nonlinear [25].

2. As a consequence of the exponential behavior and very low current densities, the effect of transistor mismatch becomes more severe than in strong inversion [26].
3. As the circuit and signal bandwidths can be very low, $1/f$ (flicker) noise becomes dominant. To suppress the $1/f$ noise, transistors are thus required to occupy large chip area, and/or a sampled-data technique, such as the auto-zeroing mechanism, needs to be applied [27].

Designing any analog function under the influence of the aforementioned issues is indeed challenging. Furthermore, to serve the requirements of miniaturized wearable and implantable medical devices, the degrees of freedom in the design are even more limited. This book thus presents a design framework that involves the following methods:

1. Keeping the circuit complexity low by utilizing a minimum number of transistors, thereby minimizing the chip area, source of noise/mismatch and power consumption.
2. Applying unity gain-negative feedback loops wherever possible to reduce sensitivity to internal disturbances (transistor mismatch in this case), as well as to enhance circuit linearity.
3. In the case that additional circuit elements are required, applying compact low-voltage circuit architectures that allow class-AB operation to enable large signal swings, thereby enhancing the circuit's dynamic range and keep static power consumption low.

Following the above design framework, this book contributes to: 1) analog sampled-data, switched-current (SI) memory cells, 2) continuous-time $G_m-C$ filters and 3) four-quadrant analog multipliers. For the two former circuit classes, we achieved more than an order of magnitude improvement in figure of merits (FoMs) in the designs [6, 7]. For the latter circuit class, the first fully class-AB, four-quadrant multiplier in subthreshold CMOS is obtained [1].

## 1.4 Materials and Methods

In this book, some well-known techniques, viz., "switched-current(SI)", "transconductance-capacitance $(Gm - C)$" and "Translinear (TL)" are investigated and developed further in order to serve biomedical signal processing in a nW-power regime.

The SI technique is traditionally considered to be compact and mismatch insensitive. For this reason, it is developed to overcome the general challenges of nanopower analog design as mentioned in the previous section.

Fundamental operation and performance limitations of a SI memory cell are investigated. Based on these findings, we propose the design of a SI memory cell that offers performance enhancement and power consumption reduction. We estimate the design efficacy by using a figure of merit (FoM) that involves relevant circuit characteristics and comparing it to other designs reported previously.

The $G_m - C$ technique is widely used for filtering biomedical signals. This technique is examined focusing on the possibility to create a nanopower continuous-time filter with large time constants in a reasonably small chip area. The concept of transistorized $G_m - C$ filters (using a transistor as a $G_m$ cell) is attractive in terms of circuit compactness and power consumption. This book develops this idea further to achieve a design methodology that leads to the chip implementation of the best-FoM, power-efficient, transistorized $G_m - C$ lowpass and bandpass filters for the audio frequency range and below.

The translinear principle is considered a powerful design tool for low-power current-mode signal processing. It exploits the nonlinear large-signal characteristic of a transistor to perform various signal processing functions from different closed-loop base-emitter junction terminals arrangements (TL loop) of BJT devices (or gate-source, gate-bulk or bulk-source terminal arrangements of MOSFETs). Therefore, large signal swings can be expected and, subsequently, a large dynamic range becomes possible. This book develops a design technique that allows class-AB operation for a subthreshold current-mode analog multiplier. This finding is useful for the design of an action potential detector that distinguishes energy of action potentials from background noise.

## 1.5  Book Organization

After the introduction in **Chapter 1**,

**Chapter 2** presents a review of analog circuit techniques widely used in low-power biomedical signal processing applications (SI, $G_m - C$, and TL). Subtheshold MOS model equations, which are employed throughout this book, are also described.

Starting from here, the book will be divided into 3 parts as follows:

### Part I Analog Sampled-Data Circuit Technique

**Chapter 3** examines the SI memory cell from a feedback point of view. Fundamental features and design considerations using MOSFETs in subthreshold are discussed.

**Chapter 4** presents the design and simulation of a class-AB SI memory cell operating in the subthreshold region. It will be shown that employing negative feedback and class-AB operation employing MOSFETs in subthreshold can help to obtain a dynamic range of 77 dB at 28 nW power consumption.

## Part II Compact Continuous-Time Filters

**Chapter 5** describes the theory, design and implementation of 'single branch filters'. A bandpass filter design methodology based on this filter structure is used to obtain the best FoM to date. Measurement results for the filter chip are also reported.

**Chapter 6** presents a lowpass filter (LPF) design based on a circuit cell called a 'follower integrator (FI)' which is a special class of single branch filters that contains a local negative feedback loop and allows for a cascade connection in a low-voltage environment. The LPF design will be discussed along with its application in ECG recording and measurement results.

## Part III Very Low-Frequency Filtering and Large-Swing Multiplication

**Chapter 7** presents a modular transconductance reduction technique dedicated to fully-integrated $G_m - C$ biomedical filters with cutoff frequencies in the range of 1–100 Hz. The chip measurement results of the $G_m - C$ lowpass filter will also be discussed.

**Chapter 8** shows a non-linear cancellation technique for realizing a class-AB four-quadrant current multiplier. This technique utilizes subthreshold class-AB transconductors and a geometric mean current splitter so that a four-quadrant multiplier with (theoretically) unlimited signal swing is obtained.

**Chapter 9** summarizes this book and discusses some outlooks.

**Appendix A** presents a peak instant detector (PID) designed for an ultra low-power analog cochlear implant speech processor.

**Appendix B** provides a detailed calculation for harmonic distortions of the class-A and class-AB current-mode sample-and-hold circuits.

# 2

# Review of Relevant Techniques
# and MOSFET Model

## 2.1 Introduction

This chapter provides a review of circuit techniques that can be useful for the design of analog ICs to be used in biomedical applications. The mentioned techniques comprise switched-current (SI), transconductance-capacitor $(G_m - C)$ and translinear (TL). The fundamental concepts and recent progresses of these technique are discussed. As the weak inversion CMOS devices are basic elements applied throughout this book, the descriptions of subthreshold MOS operation and the model used are also provided.

## 2.2 Switched-Current Technique

SI is an analog sampled-data technique recognized for its potential of being compact and insensitive to mismatch [28]. Thus it is expected to overcome the nanopower design challenges mentioned in the previous chapter. This section presents the circuit descriptions of two generations of SI memory cells. Effects of practical MOS switches and techniques to reduce the switching error are also discussed.

### 2.2.1 1st and 2nd-Generation SI Memory Cells

Figure 2.1 shows two generations of SI circuits. The 1st-generation SI memory cell [29] comprises switch $S_1$ controlled by a single phase clock signal and transistors $M_1$ and $M_2$, biased by constant current sources $I_B$ (see Figure 2.1(a)). Since $M_1$ is diode-connected, input current $I_{in}$ is converted into a voltage which will be memorized by the gate-source parasitic capacitance $c_{gs2}$ after turning off switch $S_1$. Neglecting the channel length modulation and assuming $S_1$ is an ideal switch, $I_{in}$ will be converted into $I_{out}$ for the first

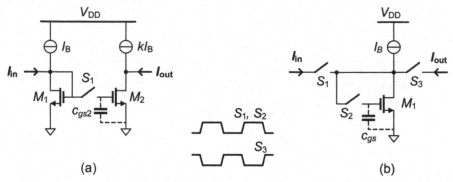

**Figure 2.1**   Switched-current memory cells: (a) 1st-generation and (b) 2nd-generation.

half period of clock signal and then, for the next half period, $I_{out}$ will hold the final value of $I_{in}$ at the end of the previous half period. We thus obtain a current-mode half delay or track-and-hold operation. It should be noticed that there is no current being switched but the gate voltage of $M_1$ is.

The true switched-current mechanism occurs in the 2nd-generation SI memory cell shown in Figure 2.1(b) [29]. Switches $S_1$ and $S_2$ are controlled by the same clock while $S_3$ is controlled by a different clock signal which is non-overlapped with the clock signal for $S_1$ and $S_2$ (see Figure 2.1). During the clock phase that $S_1$ and $S_2$ are turned on, $I_{in}$ is supplied to the diode-connected transistor creating a particular gate-source voltage. In the next clock phase, $S_1$ and $S_2$ are turned off, the gate and drain terminals of $M_1$ are disconnected and the gate-source voltage is memorized by $c_{gs}$. Simultaneously, $S_3$ is turned on supplying $I_{out}$ to the next stage. This mechanism is slightly different from the 1st-generation cell; this circuit performs current sample-and-hold operation and indeed the input and output currents of this circuit are being switched.

Regarding circuit compactness, mismatch error and power consumption, the 2nd-generation memory cell is superior to the 1st-generation one since only one transistor is used to perform both sampling and holding operations, and three switches which do not consume any static power. Therefore, the 2nd-generation SI memory cell is chosen to be developed further in this book.

## 2.2.2 Switching Error Cancellation in a SI Memory Cells

The major problem that degrades the accuracy and limits the high frequency performance of the SI memory cell is the switching error due to the non-ideal characteristics of a MOS switch viz., charge injection and clock-feedthrough effects [30]. To minimize the effects of these switching errors, several techniques have been proposed.

**Error Voltage Cancellation:** Figure 2.2(a) shows an error cancellation technique proposed for the $1^{st}$-generation memory cell [31]. An extra switch and a voltage follower/level shifter are inserted in addition to the conventional $1^{st}$-generation memory cell. With this circuit arrangement, the charge induced by switches $S_1$ will be converted into error voltages appearing at the source and gate terminals of $M_2$. Assuming capacitances $c_h$ and $c_{gs2}$ are identical, the voltage error across $c_{gs2}$ is thus minimized. Note that this technique can also be applied to the $2^{nd}$-generation memory cell. However, as the switching error is signal dependent, the error cancellation cannot be achieved completely.

**Error Current Cancellation:** Another technique applied to a $2^{nd}$-generation SI memory, called two-step sampling ($S^2I$) is shown in Figure 2.2(b) [32]. Unlike the technique mentioned above that reduces the voltage error at the memory capacitor, this technique minimizes the output current error by introducing two more sampling steps and one more memory element. The sampling period starts after turning on switches $S_1$ and $S_{1a}$. Transistor $M_p$ is acting as a current source supplying a DC current to diode connected transistor $M_n$. Then, input current $I_{in}$ flows into the drain terminal of $M_n$. After settling, $S_{1a}$ is turned off and $S_{1b}$ starts turning on while $S_1$ remains on. At this moment, $M_n$ is memorising $I_{in}$ plus switching error current ($I_a$) induced by $S_{1a}$. During this period of time, $M_p$ will track $I_a$ until $S_{1b}$ and $S_1$ are turned off. Right after switch $S_2$ is turned on, $M_p$ will store and supply $I_a$ and the error current generated by $S_{1b}$ ($I_b$) to the output. Note that $I_B$ is relatively small and less signal dependent compared to $I_a$ [32]. As $S_{1a}$ remains off, $M_n$ will also supply current $I_{in} + I_a$ to the output. By this mechanism, $I_a$ will be cancelled out and only $I_b$ remains.

**Zero-Voltage Switching:** The previous two techniques can only reduce the switching error to some extent. For high precision sample-and-hold operation, the SI memory cell with zero-voltage switching depicted in Figure 2.2(c) is required [33]. As the injected charge and clock-feedthrough charge can be made constant by fixing the voltages across the drain and source terminals of the MOS sampling switch, voltage amplifier $A_v$ is inserted for enhancing the loop gain around the sampling node. In this case, $C_H$ is used as a memory capacitor instead of parasitic capacitance $c_{gs}$ of a single transistor. During the sampling phase, the voltage across sampling switch $S_1$, will be fixed at a certain reference voltage (virtual ground) and the signal error voltage will be relayed to the gate terminal of $M_1$ instead. To the best of our knowledge, only this technique is able to suppress the switching error by making it signal-independent and allows for almost complete error cancellation in differential circuit operation. For this reason, this technique is chosen to be developed further for nanopower signal processing IC design. More detail of this work is presented in Part I of this book.

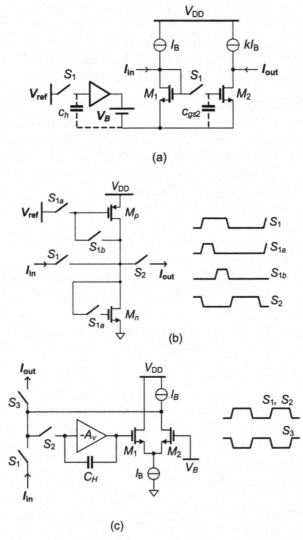

**Figure 2.2**    Switching error reduction techniques: (a) voltage error cancellation, (b) two-step sampling and (c) zero-voltage switching.

## 2.3 $G_m - C$ **Filters**

$G_m - C$ filters [34] have been applied in various moderate-linearity applications. Further research, focusing on the circuit structure and design mothodology for the $G_m - C$ filters, can be useful for nanopower biomedical signal filtering. This section provides the basic principle of $G_m - C$ filters and a survey of the recent advance in $G_m - C$ topologies.

### 2.3.1 **Basic Concept and Design Considerations**

Figure 2.3(a) shows a $G_m - C$ integrator simply formed by transconductor $G_m$ and capacitor $C$. Input voltage $V_{in}$ is converted into current $I_{out}$ according to $I_{out} = V_{in}G_m$. Then, $I_{out}$ will be integrated by $C$ resulting in output voltage $V_{out}$. An integrator is the main buiding block for synthesizing higher-order filters [34]. Figure 2.3(b) shows how the integrator can be extended to realized a 1$^{st}$-order filter. The feedback transconductor $G_{m2}$ is inserted to the output node thereby creating a lossy integrator (or 1$^{st}$-order lowpass filter (LPF)) with a transfer function of

$$H(s) = \frac{\frac{G_{m1}}{G_{m2}}}{1 + s\frac{C}{G_{m2}}}. \qquad (2.1)$$

It can be found from (2.1) that the passband gain of this filter is defined by the ratio of $G_{m1}$ and $G_{m2}$. Also, the cutoff frequency is $G_{m2}/C$. In this case, the gain and cutoff frequency can be set independently.

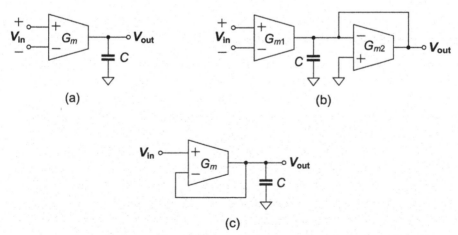

(a)

(b)

(c)

**Figure 2.3** $G_m - C$ filters. (a) Integrator, (b) 1$^{st}$-order LPF and (c) compact 1$^{st}$-order LPF.

In practice, either MOST or BJT can be used to design a $G_m$ circuit. Non-linearity of these types of transistors will affect the linearity of the filter. Thus, any $G_m$ circuit is an active element that is, not only non-linear, but also noisy. Moreover, it can be seen from (2.1) that mismatch between $G_{m1}$ and $G_{m2}$ degrades the precision of the passband gain. For these reasons, inspecting the filter topology closely can help selecting the best suitable topology for a specific application. Let's focus on circuit in Figures. 2.3(a) and 2.3(b) again. Assuming the transconductance characteristics of all of the $G_m$ circuits are identical and weakly nonlinear, the distortion produced at the outputs of the filters will depend on the voltage appearing across the input terminals of the $G_m$ circuits. For both filter circuits, we can see that the input voltage is applied directly to the input terminals of the transconductors, $G_m$ and $G_{m1}$, and thus the nonlinearities of the $G_m$ and $G_{m1}$ are fully responsible to the distortion obtained.

Figure 2.3(c) presents a first order LPF developed from the LPF in Figure 2.3(b) for $G_m = G_{m1} = G_{m2}$. In this case the passband gain is always unity (and cannot be adjusted). In term of linearity, we can consider that $V_{out}$ is connected to the inverting input of $G_m$ thereby creating a unity-gain negative feedback. For frequencies well below the cutoff frequency, $V_{out}$ will follow $V_{in}$ closely. This results in a very small voltage appearing at the differential input of the $G_m$ circuit. As a consequence, less distortion will be produced in comparison with the LPF in Figure 2.3(b). In terms of power consumption and noise, it is clear that the LPF in Figure 2.3(c) consumes less power and produces less noise than the LPF in Figure 2.3(b) for the same cutoff frequency.

## 2.3.2 Power-Efficient $G_m - C$ Filters

Based on the circuit inspection mentioned above, the idea of designing a compact LPF has been introduced in [35]. By considering a single transistor as a feedback transconductor, a differential 1st-order LPF can be realized using only two source follower circuits as shown in Figure 2.4(a). The differential structure is used here to maximize the filter's linear range.

Arising from Figure 2.4(a), the biquad section shown in Figure 2.4(b) has been developed. This circuit has been successfully employed to implement a 4th-order LPF with a 10 MHz cutoff frequency [35]. Recently, a 100 Hz LPF for biopotential recording that consumes only 15 nW quiescent power has been reported [36]. Unfortunately, inspecting this circuit further reveals that it requires a DC supply ($V_{DD}$-$V_{SS}$) of more than two gate-source voltages.

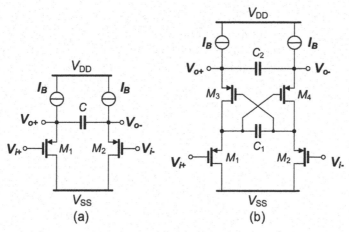

**Figure 2.4**  Transistorized $G_m - C$ filter. (a) $1^{st}$-order LPF circuit. (b) Lowpass biquad section.

Moreover, cascade connection to achieve a higher-order filter needs an even larger DC supply. Otherwise, an nMOS version of the biquad section is needed. The problem of using complementary devices in a standard CMOS process is that the bulk effect gives rise to performance degradation.

Favored by the circuit compactness, this type of $Gm - C$ circuit has been developed further in order to overcome the limitations mentioned above and to allow for application to the design of a bandpass filter. Part II of this book presents the details of this activity.

## 2.4 Translinear Circuits

The TL principle is an efficient tool for synthesizing current-mode signal processing circuits [37]. Mainly, TL circuits are designed using large-signal behaviors of MOSFET [38] and BJT (bipolar junction transistor) [37]. There-fore, the TL circuits attribute large signal swing. This section investigates the TL principle applied to BJT circuits as it can be applied directly to subthreshold CMOS circuits.

### 2.4.1 TL Principle

TL principle was first introduced based on the exponential relation between voltage and current found in BJTs and diodes [37]. According to the fact that the transconductance of a BJT varies linearly with its collector current, circuits that contain loops of base-emitter (BE) junctions allow current-mode signal processing functions to be implemented. A TL loop is characterized by an even number of junctions. The number of clockwise-oriented (CW) devices equals the number of devices with-counter clockwise (CCW) orientation. The TL principle for BJT states that "*the product of the current densities in a clockwise direction equals the product of current densities in a counter-clockwise direction*" [37]. This can be written as.

$$\prod \left(\frac{I_{Cj}}{A_j}\right)_{CW} = \prod \left(\frac{I_{Cj}}{A_j}\right)_{CCW} \qquad (2.2)$$

where $I_C$ and $A$ are the collector current and emitter area of the BJT, respectively.

Considering the circuit in Figure 2.5, it can be seen that there are four transistors connected via their BE junctions starting from $V_B$ through $Q_1$, $Q_2$, $Q_3$ and $Q_4$, and back to $V_B$ again. This arrangement forms the TL loop in which the clockwise elements are $Q_1$ and $Q_3$. $Q_2$ and $Q_4$ are the counter-clockwise elements. From the TL loop, in forms of voltage we can find that

$$V_{BE1} + V_{BE3} = V_{BE2} + V_{BE4}. \qquad (2.3)$$

Suppose that all the transistors have the same emitter area, applying the TL principle, the following relationship can be found:

$$I_{C1} \cdot I_{C3} = I_{C2} \cdot I_{C4}. \qquad (2.4)$$

### 2.4.2 Exponential and sinh Transconductors

Figure 2.6(a) shows a two-transistor TL loop with a voltage soure ($V_{id} = V_1 - V_2$), a so-called "translinear network" (TN). Summation of the voltage around the loop yields

$$V_{id} = V_{BE1} - V_{BE2}. \qquad (2.5)$$

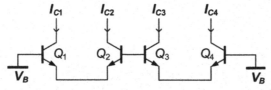

**Figure 2.5**   4-transistor translinear loop.

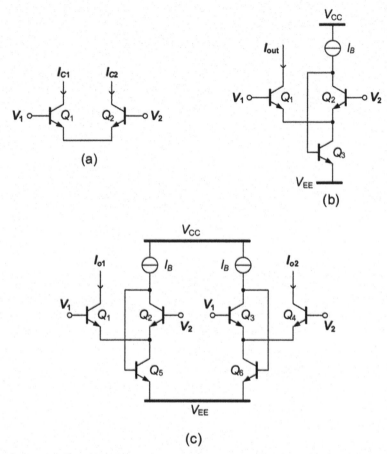

**Figure 2.6**  Translinear networks: (a) 2-transistor TL network with voltage source, (b) exponential transconductor and (c) sinh transconductor.

This relationship leads to a equation that is useful for analysing the large-signal behavior of circuits that involve this TN;

$$\frac{I_{C1}}{I_{C2}} = \exp\left(\frac{V_{id}}{V_T}\right). \tag{2.6}$$

The exponential transconductor circuit shown in Figure 2.6(b) is a good example of such a TN. Applying (2.6) and neglecting the base currents of all transistors, the voltage-to-current relationship of the circuit in Figure 2.6(b) can be found as

$$I_{\text{out}} = I_B \exp\left(\frac{V_{\text{id}}}{V_T}\right). \tag{2.7}$$

Figure 2.6(c) shows a class-AB nonlinear transconductor based on the exponential transconductor shown in Figure 2.6(b). This transconductor provides a (theoretically) unlimited output current that benefits the design of current-mode sample-and-hold and analog multiplier circuits (more details can be found in Parts I and III of this book). The voltage-to-current relationship of this transconductor is

$$I_{\text{out}} = I_{o1} - I_{o2} = I_B \left[\exp\left(\frac{V_{\text{id}}}{V_T}\right) - \exp\left(\frac{-V_{\text{id}}}{V_T}\right)\right] = 2I_B \sinh\left(\frac{V_{\text{id}}}{V_T}\right). \tag{2.8}$$

### 2.4.3 Current-Mode Analog Multiplier

Another application of TL circuits is an analog multiplier. Based on the 4-transistor TL loop shown in Figure 2.5, a current-mode four-quadrant analog multiplier can be realized as shown in Figure 2.7. Applying the TL principle to the loop containing $Q_1$ to $Q_4$ and assuming the base currents of all transistors are negligible, we have

$$(i_1 + I_B)(i_2 + I_B) = I_{C4}I_B. \tag{2.9}$$

Re-arranging (2.9) gives

$$I_{C4} = \frac{i_1 \cdot i_2}{I_B} + (i_1 + i_2) + I_B. \tag{2.10}$$

Considering the output terminal, we can see that

$$i_{\text{out}} = I_B + i_1 + i_2 - I_{C4}. \tag{2.11}$$

Substituting (2.10) into (2.11) results in

$$i_{\text{out}} = \frac{i_1 i_2}{I_B}. \tag{2.12}$$

It can be seen that a four-quadrant multiplier is achieved. However, this multiplier is operated in class-A that an input signal greater than $I_B$ cannot be applied. In Part III of this book, a class-AB four-quadrant multiplier that can handle input signals larger than the bias current will be presented.

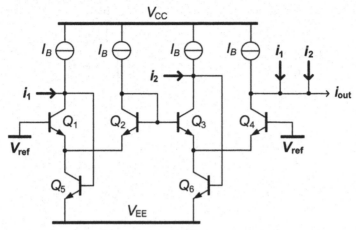

**Figure 2.7**    Class-A, current-mode, 4-quadrant analog multiplier.

## 2.5 EKV MOS Model for Low-Current Analog Design

A compact EKV model [25] developed for low-current analog circuit design is described in this section. Focusing on subthreshold operation, only relevant parameters are summarized for practical hand calculations. As the physical structures of nMOS and pMOS are intrinsically symmetrical, the bulk (or body) terminal voltage is used as the reference potential (instead of the conventional source terminal voltage) for the potential of each terminal depicted in the symbols shown in Figure 2.8. Moreover, only the nMOS device model will be described, in order to avoid repeating redundant descriptions of the dual pMOS. However, the model description for the pMOS device can be obtained by applying an opposite sign in the threshold voltage and applying the symmetric condition that

$$I_D\ (V_{\mathrm{GB}},\ V_{\mathrm{DB}},\ V_{\mathrm{SB}})\ \text{for nMOS} \equiv -I_D\ (-V_{\mathrm{GB}},\ -V_{\mathrm{DB}},\ -V_{\mathrm{SB}})\ \text{for pMOS,}$$

$$(2.13)$$

where $I_D$ is the drain current of each device, $V_{\mathrm{GB}}$, $V_{\mathrm{SB}}$ and $V_{\mathrm{DB}}$ are the gate-bulk, source-bulk and drain-bulk voltages, respectively.

### 2.5.1 Large-Signal Equations

As indicated in (2.13), the drain current is a function of all terminal voltages. For an nMOS device, the EKV model proposes a general equation for the drain current valid for all operating regions as

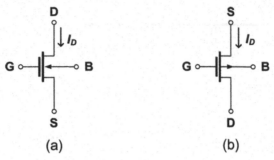

**Figure 2.8**    MOSFET 4-terminal symbols: (a) nMOS and (b) pMOS.

$$I_D = I_{Sn} \left[ \ln^2 \left( 1 + \exp \left( \frac{V_{GB} - V_{THn} - n_n V_{SB}}{n_n V_T} \right) \right) \right. $$
$$\left. - \ln^2 \left( 1 + \exp \left( \frac{V_{GB} - V_{THn} - n_n V_{DB}}{n_n V_T} \right) \right) \right] \qquad (2.14)$$

where $I_{Sn} = 2 n_n k_n (W/L) V_T^2$ is the specific current, $n_n$ is the subthreshold slope factor and $V_{THn}$ is the threshold voltage of the nMOS, while $V_T$ is the well-known thermal voltage, $k_n$ is the process transconductance parameter, and $W$ and $L$ are the channel width and length of the nMOS, respectively.

Under the subthreshold condition that $V_{GB} < V_{THn}$, the drain current behavior can be divided into different operating regions as follows.

**Weak Inversion Conduction:** This region is defined supposing that $V_{SB}$ (and $V_{DB}$) $\gg (V_{GB} - V_{THn})/n_n$. Then, (2.14) can be reduced to

$$I_D = I_{Sn} \exp \left( \frac{V_{GB} - V_{THn}}{n_n V_T} \right) \left[ \exp \left( \frac{-V_{SB}}{V_T} \right) - \exp \left( \frac{-V_{DB}}{V_T} \right) \right] \qquad (2.15)$$

It can be seen from (2.15) that $I_D$ of this region is controlled, not only by the gate voltage, but also the voltage across the drain and source terminals. In other words, the nMOS is acting as a voltage-controlled nonlinear conductor.

**Weak Inversion Reverse Saturation:** This region is found when $(V_{SB} - V_{DB}) \gg V_T$ and (2.14) can be simplified as

$$I_D = -I_{Sn} \exp \left( \frac{V_{GB} - V_{THn}}{n_n V_T} \right) \left[ \exp \left( \frac{-V_{DB}}{V_T} \right) \right] \qquad (2.16)$$

In this case, $I_D$ will flow in an opposite direction compared with the previous case and it will reach 0 when $V_{DB}$ is approximately larger than $4V_T$.

**Weak Inversion Forward Saturation:** This region is found when $(V_{DB} - V_{SB}) \gg V_T$ (in practice, $V_{DS} > 4V_T$ is sufficient). Then, (2.14) can be simplified to

$$I_D = I_{Sn} \exp\left(\frac{V_{GB} - V_{THn}}{n_n V_T}\right) \left[\exp\left(\frac{-V_{SB}}{V_T}\right)\right] \qquad (2.17)$$

For pMOS devices used in this book, the source and bulk terminals are connected together thereby avoiding the bulk effect in standard CMOS technology. By doing so, (2.17) can be simplified further and using the symmetric condition as described in (2.13), we have

$$I_D = I_{Sp} \exp\left(\frac{V_{SG} - |V_{THp}|}{n_p V_T}\right)$$

$$= I_{Sp} \exp\left(\frac{-|V_{THp}|}{n_p V_T}\right) \exp\left(\frac{V_{SG}}{n_p V_T}\right) = I_{D0} \exp\left(\frac{V_{SG}}{n_p V_T}\right) \qquad (2.18)$$

where $I_{D0} = I_{Sp} \exp\left(\frac{|V_{THp}|}{n_p V_T}\right) = I_{S0}(W/L)$ is the zero-biased current of the pMOST obtained by setting $V_{GS} = 0$ while $I_{S0}$ is defined here as the zero-biased current for a unit transistor obtained from conditions: $V_{GS} = 0$ and $W = L$. $V_{THp}$ and $n_p$ are the threshold voltage and slope factor of the pMOS device, respectively. Note that since this operating region is often used, the more appropriate term "weak inversion saturation" is used in this book to denote this region.

### 2.5.2 Small-Signal Model

Figure 2.9 shows the static (low-frequency) small-signal model of the 4-terminal transistor. Transconductances $g_{mg}$ (gate tranconductance), $g_{md}$ (drain transconductance) and $g_{ms}$ (source transconductance) contribute to the drain current variation when small variations of the respective voltages, $V_{GB}$, $V_{DB}$ and $V_{SB}$ are applied. For the weak inversion saturation nMOS, it can be found that

$$g_{mg} = \frac{\partial I_D}{\partial V_{GB}} = \frac{I_D}{n_n V_T}, \qquad (2.19)$$

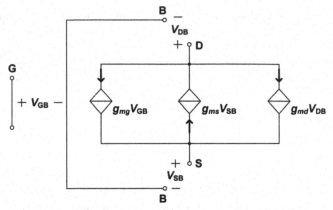

**Figure 2.9**    Static (low-frequency) small-signal equivalent circuit.

$$g_{md} = \frac{\partial I_D}{\partial V_{\text{DB}}} = 0 \ (\lambda I_D), \tag{2.20}$$

and

$$g_{ms} = \frac{\partial I_D}{\partial V_{\text{SB}}} = \frac{I_D}{V_T}, \tag{2.21}$$

Note that the channel length modulation (CLM) that leads to the drain conduction in the weak inversion saturation region is neglected. If it is included, $g_{md}$ will become as shown between brackets in (2.20), where $\lambda$ is the CLM parameter [25].

## 2.6 Conclusions

Three useful analog design techniques have been described viz., SI, $Gm - C$ and TL. The discussions of the performance limitations and advantages of the circuits based on these techniques have been given. Also a simplification of the compact EKV MOS model has been presented in order provide a few relevant nomenclatures that help the interested reader to follow the consecutive chapters conveniently. The information shown in this chapter is employed as fundamental material for the nanopower design techniques presented in the rest of this book.

# Part I

# Analog Sampled-Data Circuit Technique

# 3

# Switched-Current Technique
# in Subthreshold CMOS

## 3.1 Introduction

Processing electrical signals in the voltage domain using CMOS circuits is encountering the problem of limited voltage headroom. This results from CMOS process scaling that reduces the supply voltage and thereby forces the maximum signal voltage swing to go down [39]. To recover the signal-to-noise ratio (SNR) and the dynamic range (DR), current-mode signal processing has become attractive since the nonlinear behavior of the devices, i.e., the square and exponential laws for strong and weak inversion behaviors, respectively, provide a compressive voltage swing. A wide range of current signal swings can thus be obtained from a low supply voltage [28].

In the area of biomedical electronics that focuses on the design of portable, wearable and implantable devices, minimizing power and area consumption are major requirements. This is to keep the device sizes fit for such applications and to prolong the lifetime of the battery used. To operate circuits at very low current consumption (in the range of nA) and a supply voltage below 1 V, the CMOS devices will be forced into their weak inversion region, which creates a design difficulty in terms of noise and mismatch [40, 41]. Large chip area is required for capacitors and transistors used in a circuit to mimimize noise and mismatch. Therefore, a suitable circuit technique that can satisfy the requirements and overcome the problem of noise and mismatch is needed.

This chapter examines the feasibility to perform signal processing functions within a small silicon area to achieve SNR and DR higher than 70 dB while consuming very little electrical power. As it was introduced with the distinct feature of small area and mismatch insensitive sampled data operation, the analog current-mode technique called 'switched current' (SI) is re-examined in detail focusing on its fundamental circuit operation. Relevant effects of circuit and device non-idealities are also discussed.

25

## 3.2 Feedback Analysis of a 2nd-Generation SI Memory Cell

This section presents a feedback analysis, review of performance enhancement techniques and stability analysis of a $2^{nd}$-generation SI memory cell.

### 3.2.1 Reexamination of a 2nd-Generation SI Memory Cell

Figure 3.1(a) shows a $2^{nd}$-generation SI memory cell [28]. It comprises only one transistor biased by constant current $I_B$ and switches $S_1 - S_3$ controlled by two non-overlapping clock signals. Considering small-signal operation and including channel length modulation, the circuit in Figure 3.1(a) can be modeled as shown in Figure 3.1(b), where $R_o$ and $G_m$ represent the output resistance (output resistance of $I_B$ in parallel with that of the transistor) and transconductance factor of the transistor, respectively.

During the sampling phase ($S_1$ and $S_2$ are closed and $S_3$ is opened), the gate and drain terminals of the transistor are connected creating a feedback loop as shown in the block diagram in Figure 3.2. As one can see, the error current resulting from $I_{in} - I_f$, (where $I_{in}$ and $I_f$ represent the input and feedback currents, respectively) will flow into input impedance $Z_i = R_o(1 + sC_H R_o)^{-1}$, thereby creating voltage $V_H$ which is the input voltage of transconductor $G_m$. Voltage $V_H$ will be converted into current $I_f$ by $G_m$.

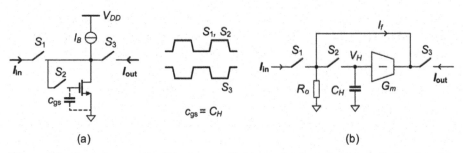

|        (a)        |        (b)        |

**Figure 3.1**   Second-generation SI memory cell. (a) circuit schematic. (b) small-signal model.

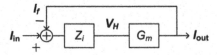

**Figure 3.2**   Feedback block diagram of the second-generation SI memory cell.

From the block diagram, the loop gain (LG) of the system can be found as

$$LG = G_m Z_i = \frac{G_m R_o}{1 + sC_H R_o}, \tag{3.1}$$

where $C_H$ equals the parasitic gate-source capacitance $c_{gs}$ of the transistor. In this case, LG equals the intrinsic gain of a single transistor which is becoming smaller in deep submicron technology [42]. The input impedance of the circuit can be also found to be

$$Z_{in} = \frac{V_H}{I_{in}} = \frac{Z_i}{1 + LG} \cong \frac{1}{G_m} \frac{1}{\left(1 + \frac{sC_H}{G_m}\right)}. \tag{3.2}$$

It can be seen from (3.1) and (3.2) that $R_o$ directly contributes to LG but affects $Z_{in}$ insignificantly. On the other hand, $R_o$ plays a role when the feedback loop is broken during the hold phase ($S_1$ and $S_2$ are opened and $S_3$ is closed). It defines the output resistance of the memory cell since the gate voltage of the transistor is held constant by the charge stored on memory capacitor $c_{gs}$.

### 3.2.2 Reconsideration of the Performance Enhancement Techniques

As we can see from the previous subsection that the operation of a 2$^{nd}$-SI memory cell is a unity-gain negative feedback system, the system's LG can be used to define the precision of the SI circuit. For this reason, the LG defined by (3.1) should be enhanced further to improve the current-mode sample and hold (CSH) operation.

There are two different approaches to enhance the LG: 1) increasing $R_o$ by cascoding transistors [43] and 2) increasing $G_m$ by cascading $G_m$ stages [3, 6, 33, 44–46]. At first glance, these two solutions seem to provide a satisfactory improvement as long as the LG is enhanced sufficiently. This is true only for the case of a continuous-time signal for which the feedback loop is always maintained. Although cascodes are used in many switched capacitor circuits, e.g., a folded-cascode opamp, for the sample and hold operation in which the feedback-loop is being switched and the swiching mechanism is performed by MOS switches, the latter solution is preferable. This is because it gives the possibility to suppress the error from charge injection and clock-feedthrough effects. As we have seen from Section 3.2, cascoding does not help regulating the voltage swing at the sampling node. The voltage at the switching node $V_H$ varies according to $I_{in}/G_m$ inducing a signal-dependent charge injection error which leads to output signal distortion [47]. This signal-dependent error is a result of the charge injection and clock-feedthrough via the gate-source and gate-drain parasitic capacitances of a practical MOS device that forms switch $S_2$ which is placed at the circuit node with signal-dependent voltage fluctuation [30]. Having a larger $Z_{in}$ due to lower $G_m$, leads to a greater $V_H$ swing and, subsequently, a more signal-dependent error is obtained. On the other hand, for a larger $G_m$, a smaller voltage swing is what we obtain from Section 3.2 and this helps the charge injection error to become less signal-dependent such that it can be possibly cancelled out by operating the CSH circuit in a differential fashion. However, to have higher $G_m$, more power consumption needs to be sacrificed.

The $G_m$ enhancement technique can be realized as shown in Figure 3.3. In Figure 3.3(a), a voltage amplifier $A_v$ is inserted in front of the $G_m$. This results in a higher effective transconductance $G_{mt} = A_v G_m$, which can be made very large. By doing so, the error current is forced to be very small by the larger LG resulting in a very small variation of $V_H$. Instead, the large voltage swing ($V_{Hlarge} = A_v V_H$) is now relayed to the output of amplifier $A_v$ which is not where the switch is placed. Therefore, the charge injection error can be

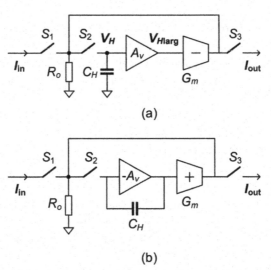

(a)

(b)

**Figure 3.3**   CSH circuit with LG enhancement with (a) grounded holding capacitor and (b) Miller holding capacitor.

considered signal-independent. To realize voltage amplifier $A_v$, another $G_m$ stage is used and unfortunately at least one additional time-constant is introduced by the parasitic resistances and capacitances of all the active elements, which may lead to instability. Pole splitting can be applied to stabilize the system by changing the location of the holding capacitor $C_H$ (which is now used as a Miller capacitor in the sampling phase) and the polarities of amplifiers $A_v$ and $G_m$ as shown in Figure 3.3(b) [33, 45]. For proper frequency compensation, the bandwidth of the CSH will be limited. This is a fundamental trade off of a low-distortion CSH circuit. Although the charge injection error of the single-ended circuit in Figure 3.3(b) can be made almost constant by the feedback technique mentioned above, the distortion still remains as long as the $G_m$ used is nonlinear.

To get rid of the charge injection error thereby minimizing distortion of the output signal, a fully differential structure, as shown in Figure 3.4, is desirable. In the case that the pair of switches $S_2$ is identical and the pair of holding capacitors $C_H$ are perfectly matched, the same amounts of constant charge injection error voltages will appear at the input terminals of the $G_m$ with the same amplitude and phase. These error voltages are therefore seen as a common-mode signal and suppressed by the common-mode rejection capability of the $G_m$. As a result, a high linearity CSH circuit is obtained [3, 44, 47, 48]. It is worth noting that even in the situation that both $G_m$ circuits are nonlinear, the complete error cancellation mentioned above can be achieved as long as capacitors $C_H$ and switches $S_2$ are identical and the former are linear, and the sampling period is sufficiently long for complete settling of $V_H$. Unfortunately, for the case that both capacitors $C_H$ are weakly nonlinear and/or switches $S_2$ are not matched perfectly, the charge injection error voltages can only be cancelled out partially. Subsequently,

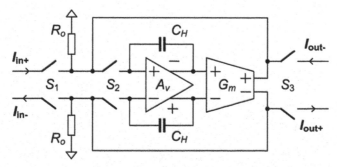

**Figure 3.4**   Fully differential CSH circuit.

output distortion will be generated from the residual input offset of $G_{m2}$. Effects of this imperfection will be discussed analytically in Section 3.3.4.

### 3.2.3 Stability and Transient Behavior

In practice, the voltage amplifier within the feedback loop of Figure 3.4 can be formed by a transconductor with high resistive loads, and the DC voltage levels at the internal nodes need to be stabilized by common-mode feedback (CMFB) circuits. Including parasitic capacitances and resistances, a more practical CSH circuit can be represented by the macro-model shown in Figure 3.5. Assuming all the circuit elements are linear, omitting the CMFB circuits and breaking the loop at the input of $G_{m2}$, the circuit can be redrawn as in Figure 3.6 to find the circuit's LG. It can be seen that the circuit is now in the form of a generic two-stage amplifier and the LG can be found to be [49]

$$\frac{V_t'}{V_t} \cong \frac{4G_{m1}G_{m2}R_1R_2\left(1 - \frac{sC_H}{G_{m1}}\right)}{s^2R_1R_2\left(C_1C_2 + C_HC_1 + C_HC_2\right) + sG_{m1}R_1R_2C_H + 1}. \tag{3.3}$$

Its open-loop unity gain frequency, poles and RHP zero can be approximated to be

$$\omega_u \cong G_{m2}/C_H, \tag{3.4}$$

$$\omega_{p2} \cong -G_{m1}/(C_1 + C_2), \tag{3.5}$$

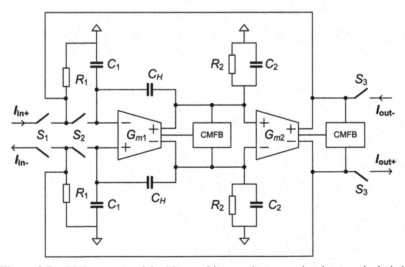

**Figure 3.5**   CSH macro-model with parasitic capacitances and resistances included.

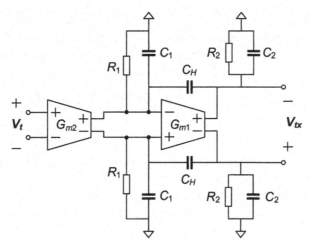

**Figure 3.6** Broken loop circuit for LG analysis.

and

$$\omega_{z1} \cong G_{m1}/C_H, \qquad (3.6)$$

respectively.

Capacitors $C_H$ are now serving two purposes: 1) the holding capacitors and, 2) pole-splitting compensation capacitors. In this case, the latter purpose defines the value of $C_H$ to maintain stability. To achieve a phase margin ($\phi_M$) of 60°, we set $\omega_{p2} \geq 2.2\omega_u$ and $C_H \ll C_1 + C_2$.

To estimate how fast a clock signal can be applied to this CSH circuit, the settling time, $t_s$ of the closed-loop response of the system in Figure 3.5 needs to be found. Within the range of an acceptable normalized output settling error ($\varepsilon$), from (3.3) we can find that [50]

$$t_s \cong \frac{2\pi C_H}{G_{m2}\sqrt{4\tan\phi_M\,(1-\tan\phi_M)}}, \qquad (3.7)$$

where the settling error is approximated as

$$\varepsilon \cong \exp(-\pi\sqrt{\tan\phi_M/(4-\tan\phi_M)}\,). \qquad (3.8)$$

Thus, we find the maximum sampling frequency of this sample and hold as $f_{s,\max} \leq 0.5t_s^{-1}$. Note that this analysis is based on the assumption that the on-resistances of all MOS switches and parasitic resistances and capacitances of the CMFB circuits are small enough to create very small time constants compared to $C_H/G_{m2}$, and it is valid for $\phi_M$ greater than 45°. Note that, in this case, $G_{m2}$ is a large-signal transconductance which is not constant but dependent on the input signal amplitude as shown in (3.19) or (3.20).

## 3.3 Design Consideration: Class-A and Class-AB

For low-voltage design, defined by $V_{DD} < 2V_{th}$ [51], there are two choices of subthreshold circuit cells to replace $G_{m2}$ in the previous section to form a CSH circuit: class-A and class-AB transconductors [6], as shown in Figures 3.7(a) and 3.7(b), respectively. Assuming all the transistors are working in weak inversion saturation ($V_{GS} < V_{th}$ and $V_{DS} > 4V_T$) [25], the large-signal characteristics of the class-A and class-AB transconductors can be expressed by

$$I_{od} = \frac{I_{o+} - I_{o-}}{2} = \frac{I_{B1}}{2} \tanh\left(\frac{V_{i+} - V_{i-}}{2n_P V_T}\right) = \frac{I_{B1}}{2} \tanh\left(\frac{V_{id}}{2n_P V_T}\right)$$

(3.9)

and

$$I_{od} = \frac{I_{o+} - I_{o-}}{2} = 2I_{B2} \sinh\left(\frac{V_{i+} - V_{i-}}{n_P V_T}\right) = 2I_{B2} \sinh\left(\frac{V_{id}}{n_P V_T}\right),$$

(3.10)

respectively, where $n_P$ is the subthreshold slope factor of the pMOSTs and $V_T = kT/q$ is the thermal voltage.

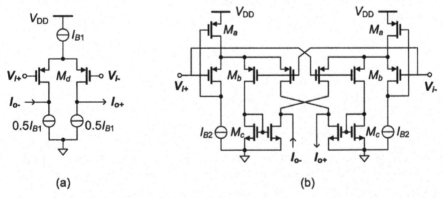

(a)                                                            (b)

**Figure 3.7**   Subthreshold transconductors. (a) Class-A. (b) Class-AB.

To design the CSH to be power efficient and to handle an input signal as large as possible, the large-signal characteristics of (3.9) and (3.10) should be neither neglected nor even approximated. In this section, we provide comparative discussions on several design aspects between class-A and class-AB CSH circuits.

### 3.3.1 Current Consumption

Considering current consumption, we divide the circuit operation into two cases: 1) static, which is defined as the situation in which there is no incoming signal, and 2) dynamic, which is the situation in which the current consumption varies with the input signal. For the class-A circuit (Figure 3.7(a)), the current consumption can be found for both situations to be

$$I_{\text{staticA}} = I_{\text{dynamicA}} = I_{B1}. \tag{3.11}$$

By contrast, the class-AB circuit (Figure 3.7(b)) allows the current to go higher than its quiescent current level for the dynamic situation. This entails a more complex circuit and hence leads to more dynamic current consumption ($I_{\text{dynamicAB}}$) as shown below [6]

$$
\begin{aligned}
I_{\text{dynamicAB}} &= 2I_{B2} + 2I_{B2} \exp\left(\frac{V_{\text{id}}}{n_p V_T}\right) + 2I_{B2} \exp\left(\frac{-V_{\text{id}}}{n_p V_T}\right) \\
&= 2I_{B2}\left(1 + 2\cosh\left(\frac{V_{\text{id}}}{n_p V_T}\right)\right).
\end{aligned} \tag{3.12}
$$

The static power consumption of this class-AB circuit can be found from (3.12) by setting $V_{\text{id}}$ to zero. This results in

$$I_{\text{staticAB}} = 6I_{B2}. \tag{3.13}$$

In order to come to a reasonable comparison between these classes of circuit operation, we use the condition that provides a static condition with the same $\omega_u$ and $\phi_M$. This condition can be satisfied by equating the small-signal transconductance gains of both circuits, i.e., $g_{mA} = g_{mAB}$.

From (3.9) and (3.10) and by considering small-signal operation, we can approximate that

$$g_{mA} = \frac{I_{B1}}{4n_P V_T} \tag{3.14}$$

and

$$g_{mAB} = \frac{2I_{B2}}{n_P V_T}. \tag{3.15}$$

For $g_{mA} = g_{mAB}$, we then have $I_{B1} = 8I_{B2}$, and this leads to

$$I_{\text{staticAB}} = 0.75 I_{\text{staticA}}. \tag{3.16}$$

From now on, we will use this condition to analyze the circuit performance.

### 3.3.2  Signal Excursion and Drivability

After setting $I_{B1} = 8I_{B2}$, let's consider (3.9) and (3.10) again. In the case that the circuits in Figures 3.7(a) and 3.7(b) are working as transconductors and that the input terminals are driven by the same differential input voltage, $V_{\text{id}}$, the output currents, $I_{\text{od}}$, for both cases are shown in Figure 3.8 ($I_{B1} = 8I_{B2} = 8$ nA). It can be seen, as expected, that for a small $V_{\text{id}}$ both circuits behave linearly giving the same tranconductance. For $V_{\text{id}} > 25$ mV, the output current of the class-A circuit starts saturating but, for the class-AB circuit, it keeps increasing exponentially. This implies that, for the CSH circuit using class-A circuitry, we can not apply an input current larger than its bias current. However, using class-AB circuitry the input current magnitude can be higher than in the case of the class-A circuit.

This argument becomes clearer when we operate the transconductors as non-linear transimpedance amplifiers based on the inverse functions of (3.9) and (3.10). Hence, we apply input current $I_{\text{id}}$ and observe the behavior of output voltage $V_{\text{od}}$ for the entire range of the varied $I_{\text{id}}$. Here, we re-define variables $I_{\text{id}} \rightarrow I_X$ and $V_{\text{id}} \rightarrow V_Y$ to be used in the inverse function

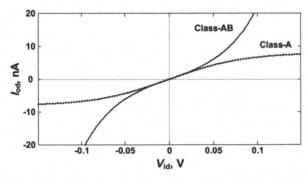

**Figure 3.8**    I-V transfer characteristics of the subthreshold transconductors.

of (3.9) and (3.10). Now, the output currents and the input voltages of the transconductors become input and output variables of the transimpedance amplifiers, respectively, as shown below.

$$V_Y = 2n_p V_T \tanh^{-1} \left( \frac{2I_X}{I_{B1}} \right) \qquad (3.17)$$

and

$$V_Y = n_p V_T \sinh^{-1} \left( \frac{I_X}{2I_{B1}} \right), \qquad (3.18)$$

for the case of class-A and class-AB, respectively.

These transfer characteristics are plotted and shown in Figure 3.9. This situation can happen ideally when negative feedback is applied and the LG is large enough to make the voltage at the input nodes constant. Then input current $I_X$ can be applied (see Figure 3.5). As $I_X$ comes close to $I_{B1}$ (8 nA), the voltage is driven to the supply voltage for a class-A circuit. This is an undesired feature for low voltage circuits in general since this large voltage excursion will push some circuit elements (transistors in this case) out of their proper operating region and eventually degrades the entire circuit performance. For class-AB, the circuit behaves in an opposite way such that, although the current goes high, the voltage can be kept low. This results from the compressive nature of class-AB operation indicated by (3.18).

Another important design parameter that should be paid attention to is the large-signal transconductance $G_m$. This parameter influences the dynamic circuit's LG. Taking the first derivatives with respect to $V_{id}$ of (3.9), (3.10) and substituting (3.17) and (3.18) into the results, we can find that

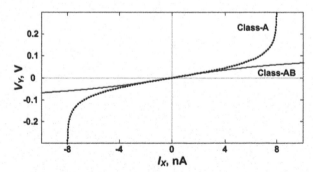

**Figure 3.9**   V-I transfer characteristics of the subthreshold transconductors.

$$G_{mA} = \frac{I_{B1}}{4n_p V_T} \mathrm{sech}^2 \left( \tanh^{-1} \left( \frac{2I_{id}}{I_{B1}} \right) \right) \qquad (3.19)$$

and

$$G_{mAB} = \frac{2I_{B2}}{n_p V_T} \cosh \left( \sinh^{-1} \left( \frac{I_{id}}{2I_{B2}} \right) \right), \qquad (3.20)$$

for class-A and class-AB circuits, respectively.

To give more insight, (3.19) and (3.20) are shown graphically in Figure 3.10. As one can see, $G_{mA}$ decreases when $|I_{id}| > 1\,\mathrm{nA}$ goes high while $G_{mAB}$ is enhanced. From these curves, we can predict that the accuracy (charge injection error cancellation) and bandwidth (see Equation 3.4) of the class-A CSH circuit will be degraded when a large $I_{id}$ is applied since the LG becomes smaller. For the class-AB CSH circuit, the accuracy and bandwidth will be enhanced to some extent. If $I_{id}$ keeps increasing the circuit will require a longer settling time and even oscillation can occur (more details will be given in the next chapter). This is a serious problem so that the maximum magnitude of $I_{id}$ needs to be identified. This will be explained in the next chapter as well.

### 3.3.3 Noise

Since both the class-A and AB CSH circuit share the same $G_{m1}$ stage, only the noise contribution from $G_{m2}$ will be considered here. The flicker noise is neglected for simplicity since it is assumed that it will be nullified by an auto-zeroing mechanism of the CSH circuit [52]. Only the current shot noise,

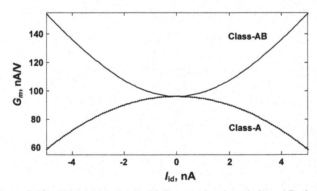

**Figure 3.10**   Transconductance of class-A circuit and class-AB circuit.

that is associated with the drain and source barriers that control the carrier concentration in the channel of the transistors, is considered here [25]. Since the spectral density of this type of noise is white, it cannot be nullified by the auto-zeroing mechanism. The output current shot noise of $G_{m2}$ will be sampled and stored on $C_H$. The stored noise will be converted into current noise again at the output during the hold phase. This sampled noise will be added to the noise generated by $G_{m2}$ during the hold phase. Due to aliasing, this type of noise becomes dominant [52].

Considering Figure 3.7 and assuming each current source to be formed by a single transistor operating in weak inversion saturation, the respective circuit schematics with their equivalent shot noise sources are shown in Figures 3.11(a) and 3.11(b) for the class-A transconductor and the half circuit of the class-AB transconductor. For the class-A case, the average output current noise power spectral density can be found to be

$$S_{ioutA}(f) = \frac{S_{in2}(f) + S_{in3}(f) + S_{in4}(f) + S_{in5}(f)}{4} \qquad (3.21)$$

$$= qI_{B1} = 8qI_{B2},$$

where $q$ represents the electron charge and $S_{ini}(f)$ is the shot noise spectral density of each transistor according to the circuit in Figure 3.11(a). The output current noise from the upper current source can be neglected at the output since it appears as a common-mode phenomenon only.

Let's now consider Figure 3.11(b). For the static condition, due to the negative feedback formed by $M_a$ and $M_b$ and its large LG, noise $i_{n6}$ does not contribute to the output. Noise sources $i_{n9}$ and $i_{n7}$ can be referred to the gate terminal of $M_b$ and relayed to the output via the transistors in the middle and the right branches. This leads to

$$S_{ioutAB}(f) = \frac{S_{ion3}(f)}{2}$$
$$= 2\left[S_{in7}(f) + S_{in9}(f)\right] +$$
$$\frac{1}{2}\left[S_{in8}(f) + S_{in10}(f) + S_{in11}(f) + S_{in12}(f)\right]$$
$$= 12qI_{B2}, \qquad (3.22)$$

where $S_{ioutAB}(f)$ is the output current noise spectral density of the class-AB transconductor and $S_{ini}(f)$ is the shot noise spectral density of each transistor corresponding to the circuit in Figure 3.11(b).

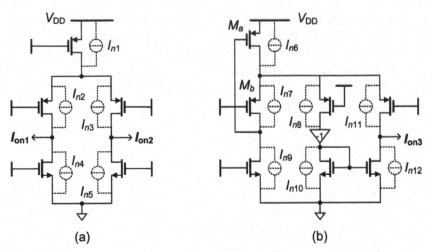

**Figure 3.11**    Transconductors with noise sources: (a) class-A and (b) class-AB half circuit.

It can be seen from a comparison of (3.21) and (3.22) that the class-AB circuit produces 50% more noise power than the class-A circuit for the static condition.

Note that (3.22) represents the output current noise power spectral density after neglecting noise generated from $G_{m1}$. In fact, both $G_{m1}$ and $G_{m2}$ contribute noise to the output and $G_{m1}$ acts as an input stage with a high voltage gain (typically more than 40 dB) and subsequently dominates the output noise power for the static situation. We know from the last section that when an input signal is applied, the drain currents of the transistors in the middle and right branches of Figure 3.11(b) ($G_{m2}$) can be many times larger than $I_{B2}$ (while there is no input current flowing into $G_{m1}$ but only its bias current) and, as a consequence, more output noise power will be generated. Therefore, for the dynamic situation with high input modulation index ($\hat{I}_{id}/I_{B2}$), the majority of output noise power will come from $G_{m2}$ instead of $G_{m1}$. However, when the input current amplitude increases beyond $I_{B2}$, the signal power increases quadratically while the noise power spectral density increases linearly, and, as a consequence, an enhanced output SNR and DR are thus obtainable for high input modulation indices.

### 3.3.4  Effects of Transistor Mismatch, Input Current Imbalance and Switching Error Cancellation

There are several causes of CSH performance degradation. They are catagorized and discussed here.

**Static Offset Voltage:** Transistor mismatch creates an offset voltage that can be modeled at the input of $G_{m2}$. As in the case of flicker noise, this offset to a large extent can be minimized by the CSH auto-zeroing mechanism.

**Input Current Imbalance:** The fully differential structure of the CSH circuit requires a balanced differential input current defined by:

$$I_{in+} = -I_{in-} = I_{id}. \tag{3.23}$$

If (3.23) can not be maintained, there will be a common-mode current flowing into the circuit. This common-mode current will be nullified by the CMFB circuit, thereby shifting either up or down the voltage at the input node corresponding to the direction of the common-mode current. For a small imbalance, this will modify the on-resistances of switches $S_1$ and $S_2$ and, as a consequence, leads to a settling time variation of the switches. For a very large imbalance, operation failure can occur.

**Switching Error Offset:** At the end of the sampling phase, the nonlinearity of, the mismatch between capacitors $C_H$, the mismatch between switches $S_2$ and an insufficient LG lead to incomplete switching error compensation. Also, this residual error can be modeled as an input offset voltage $V_{offSW}$ to $G_{m2}$, which appeares during the hold phase only and equals

$$V_{offSW} = V_{CFT+} - V_{CFT-}, \tag{3.24}$$

where $V_{CFT+}$ and $V_{CFT-}$ are error voltages induced by charge injection and clock-feedthrough effects of the MOS switches [30] appearing the non-inverting and inverting terminals of $G_{m2}$, respectively. Effects of $V_{offSW}$ will be shown for the class-A and class-AB circuits, respectively, in the following paragraph.

During the hold phase $V_{offSW}$ is added to the differential input voltage, $V_{id}$, leading to

$$I_{od} = I_{B1} \tanh \left( \frac{V_{id} + V_{offSW}}{2n_p V_T} \right), \tag{3.25}$$

and

$$I_{od} = 2I_{B2} \sinh \left( \frac{V_{id} + V_{offSW}}{n_p V_T} \right), \tag{3.26}$$

for the class-A and class-AB circuits, respectively.

Low-order harmonic distortion components can be found for the class-A circuit to be equal to

$$HD_{2A} = \frac{\tanh\left(\frac{V_{offSW}}{2n_pV_T}\right)\left(\tanh^2\left(\frac{V_{offSW}}{2n_pV_T}\right) - 1\right)}{1 - \tanh^2\left(\frac{V_{offSW}}{2n_pV_T}\right)}(MI_1), \qquad (3.27)$$

$$HD_{3A} = \tanh^2\left(\frac{V_{offSW}}{2n_pV_T}\right)(MI_1)^2, \qquad (3.28)$$

and

$$HD_{4A} = \frac{\tanh^4\left(\frac{V_{offSW}}{2n_pV_T}\right)\left(\tanh\left(\frac{V_{offSW}}{2n_pV_T}\right) - 1\right)}{1 - \tanh^2\left(\frac{V_{offSW}}{2n_pV_T}\right)}(MI_1)^3, \qquad (3.29)$$

where $MI_1 = \hat{I}_{id}/I_{B1}$ is the modulation index of the class-A transconductor and $\hat{I}_{id}$ represents the amplitude of the sinusoidal input current $I_{id}$. More detail of the distortion analysis for both class-A and class-AB circuits is presented in Appendix B.

For the case of the class-AB CSH circuit, it can be found that

$$HD_{2AB} = \frac{1}{8}\tanh\left(\frac{V_{offSW}}{n_pV_T}\right)MI_2\left(1 - \frac{1}{16}(MI_2)^2\right) \qquad (3.30)$$

and

$$HD_{4AB} = \frac{1}{256}\tanh\left(\frac{V_{offSW}}{n_pV_T}\right)(MI_2)^3, \qquad (3.31)$$

where $MI_2 = \hat{I}_{id}/I_{B2}$ is the modulation index of the class-AB transconductor.

As after expanding 3.26, a signal-dependent term of $\cosh\left(\frac{V_{id}}{n_pV_T}\right)$ which is an even function will be obtained (see Appendix B). Hence, there is no $HD_3$ for this case. For the case of $V_{offSW} = 5$ mV, $V_T = 26$ mV and $n_p = 1.6$, we can find that $HD_{2A} = 0.048$, $HD_{3A} = 0.0014$, $HD_{4A} = 6.27 \times 10^{-6}$ while $HD_{2AB} = 0.0092$ and $HD_{4AB} = 2.4 \times 10^{-4}$ for $MI_1 = MI_2 = 0.8$. In this case, $HD_{2A}$ is approximately five times greater than $HD_{2AB}$ and $HD_{3A}$ cannot be ignored while both $HD_{4A}$ and $HD_{4AB}$ are negligible.

Note that the distortion analysis here is obtained by ignoring the nonlinearity of the CMFB circuits which may further degrade the linearity of the CSH circuit. However, for comparison, this result is sufficient to support that the class-AB transconductor provides less undesired harmonic components.

### 3.3.5 Supply Noise Rejection

If we consider the class-A circuit in Figure 3.7(a), noise from supply voltage $V_{DD}$ can be seen as a current flowing through current source $I_{B1}$. The current noise will be split equally and flow through the source terminals of transistors $M_d$ appearing at the output nodes as common-mode components. Eventually, these common-mode components will be minimized by the differential operation.

For the class-AB circuit in Figure 3.7(b), assuming the circuit is perfectly symmetrical and thus comprising two identical subcircuits (each formed by the set of transistors $M_a$, $M_b$ and $M_c$, and $I_{B2}$), noise from $V_{DD}$ will flow through identical transistors $M_a$ in an equal manner. Then the noise will split equally and flow through transistors $M_b$ to the output. By assuming that the unity-gain current mirrors $M_c$ are ideal, the noise will be minimized at the output nodes while any residue noise becomes a common-mode phenomenon only.

## 3.4 Conclusions

Considering a SI memory cell from a feedback point of view reveals that a large loop gain and a fully differential circuit configuration are required to achieve a high precision sample-and-hold circuit in the current domain. It can be also seen that when transistor level circuits and non-idealities of MOS switches are considered, the class-AB transconductor is preferred to be used in the design of the current-mode sample-and-hold circuit since the internal voltage signal swing where the switches are placed can be kept almost constant while the input current can go higher than the bias current. In addition, when it comes to circuit linearity, the class-AB circuit produces lower harmonic components. Therefore, regardless of stability considerations, more than an order of magnitude larger DR can be expected for the class-AB sample-and-hold circuit despite 50% more noise contribution from the class-AB transconductor.

# 4

# A Class-AB Current-Mode
# Subthreshold SH Circuit

## 4.1 Introduction

As indicated in Chapter 3, to achieve power-efficient high-precision CSH operation, a subthreshold class-AB transconductor should be utilized. This chapter discusses several aspects of class-AB CSH circuit design. The objective is to obtain a high dynamic range of more than 70 dB from a supply voltage of less than 1 V and power consumption of less than $1\mu$W based on the two-stage fully-differential approach presented in the previous chapter. Simulation results using $0.13$-$\mu$m CMOS model parameters are also given with a comparison to designs published previously. This CSH circuit is expected to be useful in the discrete-time approach of peak detection that can benefit an analog bionic ear processor in terms of circuit complexity and type of sound produced (more detail about the peak detector circuit and bionic ear processor will be presented in Appendix A).

## 4.2 Design of a Class-AB CSH Circuit

Replacing $C_H$ by PMOS capacitors ($M_{cap}$) to reduce silicon area, as shown in Figure 4.1, the $M_{cap}$ need to be biased in strong inversion to maximize their capacitances. To do so, the input and output nodes of active element $A$ must be biased such as to accommodate the threshold voltage of $M_{cap}$. Since we would like to keep the noise power low, and we don't need a high current driving capability for this stage but rather a high voltage gain, the class-A folded-cascode transconductor shown in Figure 4.2(a) is chosen to realize element $A$. Its common-mode output voltage can be controlled by the CMFB1 circuit shown in Figure 4.2(b), where $k = 0.05$ is a scaling factor to save current consumption. The class-AB circuit in Figure 4.3(a) is used for active element $AB$ and its CMFB2 circuit is shown Figure 4.3(b). The bias current in this case

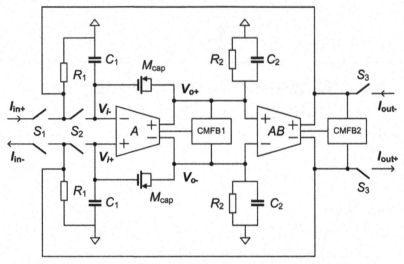

**Figure 4.1**    CSH macro-model with MOS capacitors and parasitic impedances included.

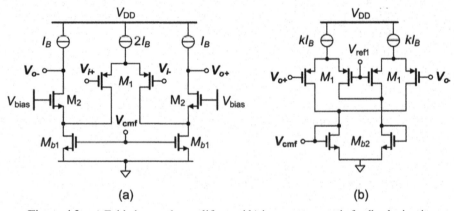

**Figure 4.2**    a) Folded-cascode amplifier and b) its common-mode feedback circuit.

is not scaled down since $I_{B2}$ is set low already ($I_{B2} = 0.4$ nA) to minimize noise and satisfy the stability condition, and by scaling it down further it may become difficult to make it precise.

## 4.2.1  Bias Condition

To keep all transistors working in weak inversion saturation the following bias conditions are set,

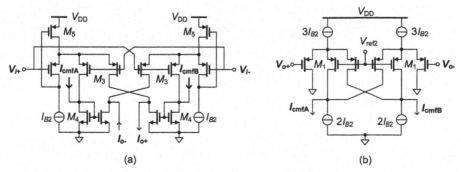

**Figure 4.3** a) Subthreshold class-AB transconductor and b) its common-mode feedback circuit.

$$V_{\text{ref2}} \cong 4V_T \text{ and } V_{\text{ref1}} \cong V_{\text{ref2}} + |V_{tp}|, \tag{4.1}$$

and

$$V_{DD} \cong V_{\text{ref1}} + V_{SG3} + 4V_T + V_{\text{swing}}, \tag{4.2}$$

where $V_{\text{swing}}$ is the internal voltage swing that follows from the relationship of (3.18).

To satisfy the desire for sufficient phase margin, $\phi_M \geq 60°$, $\omega_{p2} \geq 2.2\omega_u$ must be arranged. In order to fulfill this condition, the bias currents are set to

$$I_B = 25I_{B2}. \tag{4.3}$$

This leads to a total current consumption (excluding bias circuit) of

$$I_{Btotal} = 4.1I_B + 12I_{B2} = 114.5I_{B2}. \tag{4.4}$$

## 4.2.2 Input Current Limitation and Settling Behavior

As mentioned in the previous chapter, a $60°$ phase margin cannot be maintained for the entire range of $I_{id}$. As indicated by (3.15), $G_{mAB}$ changes according to $I_{id}$ and this leads to circuit instability for large amplitudes of $I_{id}$. We set the safety limit at a $\phi_M \cong 45°$, for which $\omega_u = \omega_{p2}$. Hence, the maximum $I_{id}$ that we can apply within this safety limit can be found as

$$I_{\text{idmax}} \cong 2I_{B2} \sinh\left(\cosh^{-1}\left(3.125\frac{C_H}{C_1 + C_2}\right)\right). \tag{4.5}$$

For $I_{id}$ larger than $I_{\text{idmax}}$, the phase margin will become smaller than $45°$.

Figure 4.4 shows a MATLAB simulation plot of $\phi_M$ versus input current amplitude, $\hat{I}_{id}$, for the following realistic parameters, $I_B = 10$ nA, $I_{B2} = 0.4$ nA, $n_p = 1.6$, $V_T = 26$ mV, $R_1 = 400$ MΩ, $R_2 = 120$ MΩ, $C_1 = 0.2$ pF, $C_2 = 0.32$ pF and $C_H = 0.25$ pF. It can be seen that for input amplitudes greater than 0.5 nA $\phi_M$ decreases rapidly.

The settling time, $t_s$, of this closed-loop system behaves consistently with $\phi_M$. Figure 4.5 shows a plot of $t_s$ versus $\hat{I}_{id}$ with $\varepsilon = 0.02$. The system response goes from overdamped to critically damped and $t_s$ decreases when $\hat{I}_{id}$ increases. For $\hat{I}_{id}$ slightly greater than 1 nA, the system response moves to the underdamped case and a ripple on $t_s$ occurs [53]. Finally, $t_s$ goes up rapidly as $\hat{I}_{id}$ approaches 10 nA since the system enters the undamped situation. This implies that the maximum sampling frequency of this CSH circuit depends on $\hat{I}_{id}$ and, in this particular example, the sampling interval should be longer than 20 $\mu$s to cover $\hat{I}_{id}$ from 0.1 nA up to 5 nA. Also for higher amplitudes (5 nA $< \hat{I}_{id} <$ 10 nA), the required sampling period rises rapidly and reaches 0.1 ms at $\hat{I}_{id} = 10$ nA.

## 4.3  Circuit Simulations

The CSH circuit has been designed and simulated in Cadence Spectre-RF™ using TSMC 0.13-$\mu$m CMOS process parameters. Transistor sizes are shown in Table 4.1. $V_{DD} = 0.6$ V, $V_{ref1} = 0.42$ V, $V_{ref2} = 0.1$ V and $C_H = 0.25$ pF.

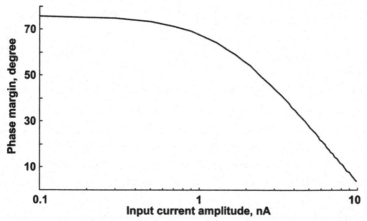

**Figure 4.4**  Simulated phase margin versus input current amplitude for $I_B = 10$ nA, $I_{B2} = $ 0.4 nA, $n_p = 1.6$, $V_T = 26$ mV, $R_1 = 400$ MΩ, $R_2 = 120$ MΩ, $C_1 = 0.2$ pF, $C_2 = 0.32$ pF and $C_H = 0.25$ pF.

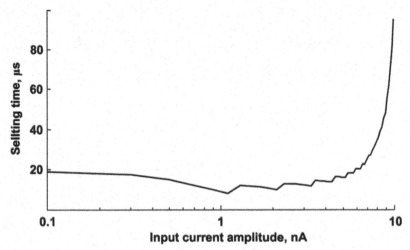

**Figure 4.5** Simulated settling time versus input current amplitude for $I_B$ = 10 nA, $I_{B2}$ = 0.4 nA, $n_p$ = 1.6, $V_T$ = 26 mV, $R_1$ = 400 MΩ, $R_2$ = 120 MΩ, $C_1$ = 0.2 pF, $C_2$ = 0.32 pF and $C_H$ = 0.25 pF.

**Table 4.1**  Transistor dimensions

| MOSFET | $W$ [$\mu$m] | $L$ [$\mu$m] |
|---|---|---|
| $M_1$ | 2 | 0.5 |
| $M_2$ | 1.5 | 0.5 |
| $M_3$ | 0.25 | 1.5 |
| $M_5$ | 0.5 | 1 |
| $M_{b1}$ | 20 | 1 |
| $M_{b2}, M_4$ | 1 | 1 |
| $M_{cap}$(0.25 pF) | 10 | 10 |

Biasing currents $I_B$ = 10 nA and $I_{B2}$ = 0.4 nA are set for $G_{m1}$ and $G_{m2}$, respectively. All switches are realized as nMOSTs with a threshold voltage of $V_{tn} \cong 0.3$ V and driven by clock signals switching between $V_{DD}$ and ground. The dimensions of the switches are identical and chosen to be as small as the process allows to minimize charge-injection and clock-feedthrough effects ($W$= 0.15 $\mu$m and $L$ = 0.13 $\mu$m). The quiescent power consumption of the entire circuit equals 27.5 nW.

Figure 4.6 shows the transient input and output currents for an input amplitude and frequency of 10 nA and 1 kHz when the CSH is sampled by a 10 kS/s clock signal with a rise and fall time of 50 ns. The large glitches appearing at the beginning of the hold phase are induced by a sudden change of

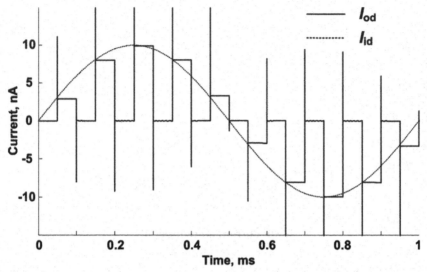

**Figure 4.6**   Transient input and output currents: $I_{id} = \hat{I}_{id}\sin(2000\pi t)$, $\hat{I}_{id} = 10$ nA and $f_s = 10$ kS/s.

the CSH circuit's output resistance as a consequence of the discontinuity of the $LG$. The non-overlapping clock transition (from switching off $S_2$ to switching on $S_3$) allows large transient voltages (products of the held currents and the large output resistances) to be produced at the output terminals of $G_{m2}$. It leads to large voltage differences across switches $S_3$ (assuming they are loaded by a similar CSH circuit having a fixed input voltage). Switching on $S_3$ will bring down the high voltages to the voltages at the input nodes of the next stage. This process happens across the drain–source parasitic capacitances of $S_3$ for a very short period of time, when the transient currents flowing through the output in addition to the desired output current are generated. This mechanism not only produces the glitches but also deteriorates the circuit's linearity. At the moment that the output voltages suddenly go low before completely closing $S_3$, small charges (fed through the parasitic gate–drain capacitances of $M_3$) will be added to $C_H$ giving a held voltage error [29]. However, there are two ways to reduce these glitches, thereby enhancing the circuit's linearity: 1) trying to eliminate the non-overlapping moment by employing a special clock scheme [54], 2) creating low impedance output nodes by introducing current followers at the output terminals of $G_{m2}$. This can be done at circuit level by cascoding output transistor $M_3$ [43]. However, since the linearity of the CSH is not severely degraded, we have not adopted either of these solutions for our design.

The internal differential voltage swings at the input of $G_{m2}$ are shown in Figure 4.7. For a 1 kHz sinusoidal input current with an amplitude of 2 nA, the CSH circuit responds slowly since $G_{m2}$ and $\omega_u$ are low and there is no ringing during the entire cycle. For the case of a higher input current amplitude (10 nA), the ringing appears when $V_{id}$ reaches 0.09 V. This is because $G_{m2}$ is enhanced according to (3.20) and as $\omega_u$ moves closer to $\omega_{p2}$: the phase margin and settling time of the CSH circuit are degraded.

Noise and linearity performance were verified using periodic steady state (PSS) and periodic noise (PNOISE) analyses for 12 harmonics. A 1 kHz sinusiodal input signal with its amplitude varying from 40 pA to 11.5 nA was applied with a 20 kS/s sampling rate. Figure 4.8 shows the output noise power integrated from 1 Hz to 10 kHz as a function of MI ($\hat{I}_{id}/I_{B2}$). It can be seen that the noise power remains constant in the range of $0.1 < \mathrm{MI} < 1$. For MI higher than 1, the noise increases. This is in line with what we predicted in the previous chapter, namely that the input current modulates the drain currents of the transistors in the class-AB transconductor, thereby creating more shot noise.

The spurious-free dynamic range (SFDR), SNR and signal to noise plus distortion ratio (SNDR) are plotted and shown in Figure 4.9. From this plot,

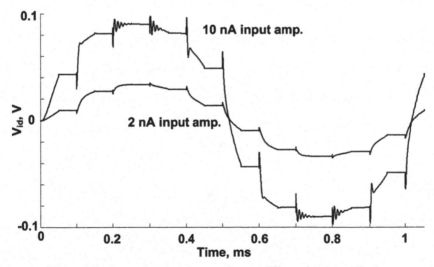

**Figure 4.7** Internal node voltage swings at different input amplitudes.

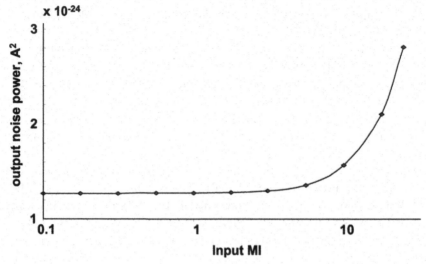

**Figure 4.8**   Integrated noise power from 1 Hz-10 kHz as a function of the modulation index (MI).

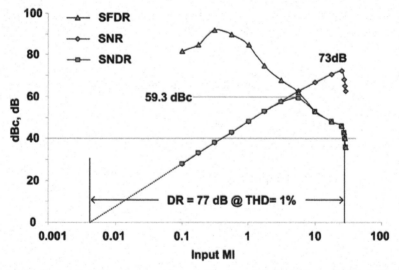

**Figure 4.9**   Spectral performance metrics as a function of the modulation index (MI). DR of 77 dB is found from the distance between MI that starts to bring the signal above the noise floor (SNR = 0 dB) and MI that brings SFDR down to 40 dB.

a DR (measured up to a 40 dB SFDR corresponding to a total harmonic distortion, THD, of 1%) of 77 dB is obtained and an SNDR of 59.3 dB can be achieved at a 2.25 nA input amplitude. This leads to an effective number of bits of

$$\text{ENOB} = \frac{\text{SNDR} - 1.76}{6.02} = 9.6\,\text{bits}. \tag{4.6}$$

Therefore, a figure of merit that embraces the effects of distortion, sampling speed, and power consumption, of

$$\text{FoM} = \frac{P}{f_s \cdot 2^{\text{ENOB}}} = 1.9\,\text{nW/MHz} \tag{4.7}$$

is obtained, where $P$ represents the average power consumption and $f_s$ is the sampling rate.

To see the effect of transistor mismatch on the circuit linearity, a Monte-Carlo transient simulation using a 976.6 Hz, 10 nA amplitude sinusoidal $I_{\text{id}}$ (corresponding to MI = 25, which is the maximum amplitude that can be applied before oscillation, see Figure 4.9) and $f_s$ = 20 kS/s has been done. The results are shown in Figure 4.10. For 100 runs, a mean value of the THD of −45 dB is obtained with a standard deviation of 3.92 dB.

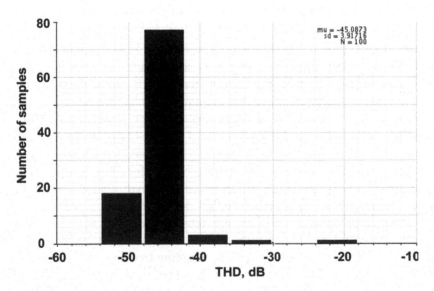

**Figure 4.10**    Monte-Carlo simulation of the THD for an input MI of 25.

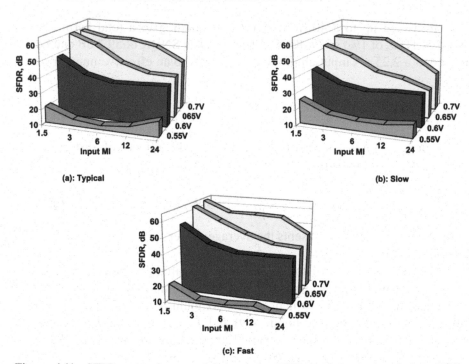

(a): Typical

(b): Slow

(c): Fast

**Figure 4.11**   SFDR versus input modulation index PVT simulations: a) typical at 40°C; b) slow at 80°C; and c) fast at 0°C cases.

Figure 4.11 shows the CSH circuit's simulated SFDR versus MI for extreme process, temperature and supply voltage conditions with the same setup as used for the above Monte-Carlo simulation. It can be seen that the minimum operating supply voltage that the circuit can handle is 0.6 V. From this supply voltage, running slow transistors at a high temperature (80°C) gives the poorest results. For MI greater than 1.5, the SFDR falls from 40 dB to around 30 dB. For higher supply voltages, better linearity is obtained for all process and temperature corners. The results obtained from these process and temperature corners are not much different. Further investigation is required to know whether process or temperature, has more influence on SFDR.

A performance comparison with previously reported CSH circuits is presented in Table 4.2. In addition to the simulations mentioned above, we also tested the CSH circuit for higher input and sampling frequencies (50 kHz and 1 MS/s, respectively). To do so, the bias current levels were changed to $I_{B1} = 25I_{B2} = 500$ nA. To handle the larger gate-source voltages of all transistors, we increased the supply voltage to $V_{DD} = 0.8$ V. The results are summarized

**Table 4.2**  Performance comparison

| Reference | [41]* | [55]* | [56]** | [57]** | This Work* | |
|---|---|---|---|---|---|---|
| Tech. [$\mu$m] | 0.35[†] | 0.18[††] | 0.35 | 0.35 | 0.13 | 0.13 |
| $V_{DD}$[V] | 1 | 3.3 | 1 | 2 | 0.6 | 0.8 |
| Static $P$ [W] | 5.8 $\mu$ | 2.6 $\mu$ | 6 m | 3 m | 6 m | 28 n | 1.84 $\mu$ |
| $f_s$ [MS/s] | 1 | 1 | 13 | 35 | 100 | 0.02 | 1 |
| $SNR_{max}$ [dB] | | 60 | 56 | 60 | 73 | 72 |
| ENOB [bit] | - | - | - | 9.6 | 9.6 | 8.7 |
| DR [dB] | 73 | 68.2 | 60 | 56 | - | 77 | 76 |
| THD [dB] @ $f_{in}$ [Hz], | -41@ na | -38.6@ na | -66@1.3 M | -55@1.32 M | -77@1 M | -40@1 k | -40@50 k |
| MI | 1 | 1 | 0.98 | 0.9 | 0.9 | 27 | 22 |

*simulation, **measurement, [†] cascoded SI cell, [††]S$^2$I cell.

in the last right column. At this bias point, transistors that form $G_{m1}$ enter moderate inversion for both static and dynamic situations. For the dynamic situation with high input modulation index, some transistors that form $G_{m2}$ will be forced into moderate inversion as well. As a consequence, parasitic capacitances $C_1$s become bigger and $C_2$s change dynamically according to the input amplitude. This affects the dynamic stability condition and results in a reduced allowable signal swing. The DR obtained becomes 2 dB less than the low-power, low-frequency operation purely based on weak inversion operation. In terms of linearity, THD better than –40 dB is obtained when MI is lower than 22 while the same level of THD can be achieved for MI up to 27 in the lower-power, low-frequency case. In comparison with other designs, as can be deduced from the $\text{SNR}_{max}$ of [55–57], class-A operation provides us at most 60 dB of dynamic range. As we can see from [41] (weak inversion class-AB SI memory cells) and this work, to reach higher than 70 dB DR, class-AB operation is required.

## 4.4 Conclusions

The design of a subthreshold class-AB CSH circuit has been presented. Benefitting from negative feedback and the exponential behavior of the transistors in weak inversion, the proposed CSH circuit can be operated from a 0.6 V supply and consumes 28 nW quiescent power. In addition to that, SNR and DR of higher than 70 dB, and a FoM of 1.9 nW/MHz are obtained. Monte-Carlo and corner simulations also confirm that a good linearity of the circuit can be maintained when realistic mismatch, process, voltage and temperature variations are taken into account.

# Part II

# Compact Continuous-Time Filters

# 5

# Nanopower BPF Using Single-Branch Biquads

## 5.1 Introduction

Analog filters are indispensable circuit building blocks in electronic systems. They separate desired signals from other signals and noise by making use of differences in their energy spectra. In order to be able to compare various filters, a figure of merit (FoM) that combines several circuit or signal parameters to a single number is often helpful. Adapting the concept of minimum possible energy per cycle and per frequency pole [58] to the design of a bandpass filter (BPF) circuit, a FoM can be defined as

$$\text{FoM}_{\text{BP}} = \frac{P}{N \cdot f_C \cdot \text{DR}}, \tag{5.1}$$

where $P$, $N$, $f_C$ and DR are the power consumption, filter order, center frequency and dynamic range of the filter, respectively. It is clear from (5.1) that the cost (numerator) over the performance (denominator) should be as low as possible. As it is commonly known that $P$ is the product of current and voltage, and DR is the ratio of maximum signal power (limited by distortion of the filter) over the minimum signal power (defined by the noise floor of the filter) that can be applied to the filter, to enhance the FoM, the following conditions should be met:

1. The filter circuit should contain the least number of current branches and operate from the lowest possible supply voltage ($V_{\text{DD}}$) for a given $N$, $f_C$ and DR.
2. The filter topology should contain a minimum number of active (noisy) elements per time constant.

For the FoM of biomedical BPF designs that have a center frequency in the audio range (and below) and a power consumption less than 1 $\mu$W [59–63],

$V_{DD}$ has been added into the numerator of (5.1) as [60]

$$\text{FoM}'_{BP} = \frac{P \cdot V_{DD}}{N \cdot f_C \cdot \text{DR}}. \tag{5.2}$$

As a consequence, $V_{DD}$ is accounted for twice and becomes the most important factor in this modified definition. Although it has been commonly used in recently reported BPFs [61–63] as reducing $V_{DD}$ is considered a virtue, the fundamental basis for this modified FoM is questionable. In this work, we therefore consider the definition of (5.1) [58] instead of the modified FoM introduced in [60].

Figure 5.1 shows a plot of FoM$_{BP}$ versus $V_{DD}$ using (5.1) for various biomedical BPFs collected from 2003 to 2012 [59–63]. The second best BPF is the design presented in [59]. Its topology satisfies the first condition for enhancing FoM (more details on the filter topology will be given in Section 5.3). The worst FoM belongs to the BPF introduced in [61]. This is because the design of [61] uses a state-space filter topology that requires many $G_m$ cells to realize both feedforward and feedback filter coefficients and integrators. Obviously, this violates condition No. 1. The BPFs introduced in [60] and [63] use very similar filter topologies (based on element substitution of a passive $LC$ ladder prototype) and provide almost the same FoMs ($2.13 \times 10^{-16}$ and $2.15 \times 10^{-16}$, respectively). It should be noted that $LC$

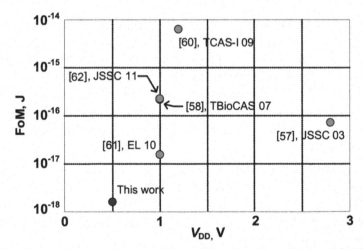

**Figure 5.1** Figure of merit comparison of $G_m - C$ bandpass filters developed for biomedical applications collected over 2003–2012, and this work.

ladder filter topologies possess an advantage of low sensititvity to parameter variations that is superior to other type of filters. This feature is not accounted for in the definition of (5.1).

These numbers show a considerable FoM improvement with respect to the design of [61], but are still worse than that of [59]. The reason for this is that substituting one floating inductor in $G_m - C$ filters requires 4 $G_m$ cells and a grounded capacitor. Although the number of active elements is less than required in the state-space filter of [61], this is not in line with condition 1, either. A significant improvement can be seen for the filter presented in [62], for which almost an order of magnitude improvement with respect to the filter of [59] is achieved. This design uses the same filter topology as used in [59], and adopts a compact, power efficient $Gm - C$ biquad structure at transistor level that requires only two branches of current consumption from [35]. The conditions for FoM enhancement have almost been fulfilled in this design. However, since the filter topologies used in [59] and [62] require a voltage follower circuit to reduce loading effects, and the transistor circuit of the biquad section used in [35] and [62] cannot be operated at a very low $V_{DD}$, there is still a possibility for the FoM to be enhanced further.

In this chapter, we develop the $G_m - C$ BPF topology further to achieve a significant FoM improvement within the context of low frequency integrated filters for biomedical applications. In this context, very large resistors would occupy a large chip area and would be severely limited in bandwidth and should be avoided. We propose:

- A new macro-model: a low-voltage $2^{nd}$-order filter topology that has a minimum number of active (noisy) elements and can be cascaded without the expense of a voltage buffer.
- A new transistor-level circuit: a low-voltage, power-efficient, single branch circuit structure using a single transistor as a $G_m$ cell that can be fit into the filter topology mentioned above.

By doing so, a $4^{th}$-order BPF with FoM improvement can be successfully realized. Measurement results of the proposed BPF fabricated in AMS 0.18-$\mu$m CMOS technology confirm our concept.

It should also be noted that there are two possible disadvantages of this method: 1) the filter's linearity is limited, resulting in a maximum signal swing of a few milli-volts, and 2) the BPF's quality factor is limited to the maximum value of 0.5. However, these limitations do not prevent this filter from being applied in various biomedical applications, such as, e.g., a filter for cochlear implant speech processing [59, 64, 65]. This chapter is developed from [7] by adding a more detailed performance analysis and measurement results.

In the next section, the idea of realizing filtering functions from a single branch circuit structure and their performance will be discussed. The proposed BPF design, including the details of the filter topology selection, transistor level architectures, linearity, common-mode behavior and design methodology will be presented in Section 5.3. Section 5.5 presents measurement results and a detailed comparison with previously published designs. Conclusions will be given in the last section.

## 5.2 Single Branch Filters

To achieve a compact power efficient filter circuit structure, this section explores the feasibility of realizing continuous-time filters from circuit structures that contain a minimum number of transistors. Moreover, the $V_{DD}$ required for the filter circuits is less than two gate-source voltages ($V_{GS}$) plus one saturation voltage ($V_{DSsat}$) and the filter employs only one branch of bias current.

### 5.2.1 Filter Topology Using Feedback Transconductors

For a MOSFET that is biased in weak inversion saturation, when the bulk and source terminals are connected to each other to minimize the bulk effect, the symbol of a four-terminal device as shown in 5.2(a) can be reduced to become three-terminal as shown in Figure 5.2(b). A differential $g_m$ cell connected in negative feedback fashion (as shown in Figure 5.2) can be obtained from the small-signal operation of the transistor.

Since the bulk effect, also known as body effect, is unwanted, a pMOS device in an n-well CMOS process that allows the source to be conected

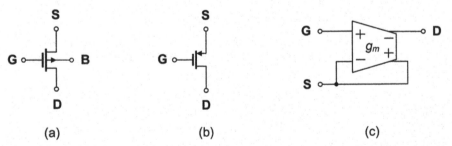

(a)                              (b)                              (c)

**Figure 5.2**   Single transistor (a) four-terminal symbol, (b) three-terminal symbol and (c) its macro-model.

to the bulk is preferred in this design. Biased by DC drain-source current $I_B$, the transconductance of the transistor with zero $V_{BS}$ (bulk and source terminals connected together) and neglecting channel length modulation is given by [25]

$$g_m = \frac{I_B}{n_p V_T}, \qquad (5.3)$$

where $n_p$ and $V_T$ represent the subthreshold slope factor of the pMOS device [25] and the thermal voltage, respectively. Using this macro-model, several filter topologies can be found from circuit structures having a single bias current branch.

Figure 5.3 shows possible realizations of single-branch $G_m - C$ filters. A lowpass (LP) filter can be obtained from the circuits in Figures 5.3(a) and 5.3(b), a highpass (HP) filter (from the former) and a bandpass (BP) filter (from the latter).

The circuits in Figure 5.3 are all formed by a cascode connection of transistors $M_1$ and $M_2$ and capacitors between source and AC ground terminals. Both transistors share the same bias current $I_B$. Replacing $M_1$ and $M_2$ by the macro-model of Figure 5.2 results in the small-signal macro-model shown in Figure 5.4. Assuming that the values of $C_1$ and $C_2$ are much greater than

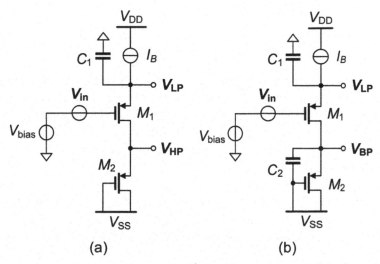

(a)                               (b)

**Figure 5.3** Single-branch filter circuits. (a) $1^{st}$-order LP and HP filters. (b) $1^{st}$-order LP and $2^{nd}$-order BP filters.

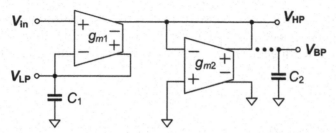

**Figure 5.4**   Single-branch filter small-signal models.

parasitic gate-source and gate-drain capacitances $c_{gs}$ and $c_{gd}$ of $M_1$ and $M_2$, we can show that:

$$H_{LP}(s) = \frac{V_{LP}(s)}{V_{in}(s)} = \frac{1}{1 + s\frac{C_1}{g_{m1}}}. \tag{5.4}$$

Without capacitor $C_2$, we obtain the following HP transfer function

$$H_{HP}(s) = \frac{V_{HP}(s)}{V_{in}(s)} = \frac{-s\frac{C_1}{g_{s2}}}{1 + s\frac{C_1}{g_{m1}}}. \tag{5.5}$$

By adding $C_2$ to the former HP output node, the following BP transfer function can be achieved:

$$H_{BP}(s) = \frac{V_{BP}(s)}{V_{in}(s)} = \frac{-s\frac{C_1}{g_{s2}}}{\left(1 + s\frac{C_1}{g_{m1}}\right)\left(1 + s\frac{C_2}{g_{s2}}\right)}. \tag{5.6}$$

It can be seen that there are three filter circuits obtained from this single branch structure. Apart from the potential for low current consumption, another advantage of these filters is that they feature a high impedance input node (being the gate terminal of $M_1$). As a result, there is no severe loading effect for cascade connections of these filters assuming that the MOS parasitics $c_{gs1}$ and $c_{gd1}$ are sufficiently small.

Since the filters are operating in weak inversion saturation and the cutoff or center frequency can be adjusted by the value of $g_m$ and $g_s$, which are proportional to $I_B$ in weak inversion saturation, a wide tuning range of the filter's cutoff frequency via controlling $I_B$ can be expected.

## 5.2.2 Supply Voltage Requirement and Current Consumption

Considering the circuits in Figure 5.3(b) in conjunction with the transfer function of (5.6), a current consumption of $0.5I_B$ per filter pole is found.

To create proper bias points in weak inversion saturation, the supply voltage $V_{DD}$ and common-mode level $V_{CM}$ should be considered. For the stacked circuit shown in Figure 5.3 and setting $V_{SS} = 0$ V, the supply voltage required must be at least $V_{DD} = V_{inpp} + V_{SG2} + 2V_{SDsat}$ (assuming that $I_B$ requires similar $V_{SDsat}$ as $M_1$). To satisfy the condition of weak inversion saturation that $V_{Dsat} \cong 4V_T$, this can be re-arranged to

$$V_{DD} \cong V_{inpp} + V_T \left( 8 + n_p \ln \left( \frac{I_B}{I_{D0}} \right) \right), \tag{5.7}$$

where $I_{D0}$ and $V_{inpp}$ are the zero-biased current of the transistors and input (peak-to-peak) voltage swing, respectively. It can be seen from (5.7) that there is a fixed term of $8V_T \cong 200$ mV and a bias current related term. The latter term is directly related to the cutoff frequency of the filter. In the case of cascading HP and BP filters, $V_{CM}$ of each stage should equal

$$V_{CM} = V_{SS} + V_{SG2}, \tag{5.8}$$

to maintain the same signal swing range for all cascaded stages.

This condition on $V_{CM}$ creates a new requirement for $V_{DD}$ to be at least $2V_{SG} + V_{SDsat}$ (assuming $V_{SG1} = V_{SG2}$ and $V_{Dsat} \cong 4V_T$ ), or in a form similar to (5.7);

$$V_{DD} \cong V_{inpp} + 2V_T \left( 2 + n_p \ln \left( \frac{I_B}{I_{D0}} \right) \right). \tag{5.9}$$

In this extremely low-power design context, an $I_B$ in the range of 0.1 to 10 nA is used to accommodate cutoff frequencies ranging from 100 Hz to 10 kHz. Therefore, either (5.7) or (5.9) can be higher, and the highest one is the minimum $V_{DD}$ required for the cascaded stage. For the 0.18-$\mu$m CMOS technology used in this work, $I_{D0} \cong 230$ pA and $n_p \cong 1.6$ is obtained for $W/L = 10$. Figure 5.5 shows a detail of the required $V_{DD}$ for $V_{inpp} = 25$ mV and different values of $I_B$ according to (5.7) and (5.9). The grey line indicates the level of $V_{DD}$ required. For $I_B$ less than 3 nA, (5.8) defines $V_{DD}$ which can be set as low as 0.2 V at $I_B = 0.1$ nA. For $I_B$ greater than 3 nA, $V_{DD}$ is defined by (5.9). From this plot, $V_{DD} = 0.5$ V is confirmed to be sufficient for the whole range of $I_B$ (from 0.1 nA to 10 nA).

On the other hand, any cascade connection that uses the LP filter increases the required $V_{DD}$ as the output is taken from the source terminal and the input is applied at the gate terminal (unless a complementary (nMOS) version of the single-branch LP filter is applied [35]). This will make either the required $V_{DD}$ eventually exceed the available supply voltage or the filter suffers from

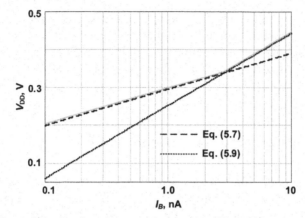

**Figure 5.5**   Supply voltage requirement for different bias currents.

the body effect. Another way to solve this problem is using a level shifter as an interface block to shift the source voltage of $M_1$ down before cascading the next stage of LPF, thereby maintaining the same required $V_{DD}$. This, however, leads to more power consumption and more noise contribution. For this reason, the LP filter will no longer be considered.

## 5.2.3 Noise

Figure 5.6(a) shows a single-branch BPF biquad circuit and its noise sources. In practice, $I_B$ can be formed by a single transistor $M_B$ biased by $V_B$. The transistor's weak inversion noise behavior will be dominated by its shot noise assuming that, for simplicity, $V_B$ is noiseless, each transistor is sized large enough and its drain current is low enough to keep the $1/f$ noise corner frequency below the frequency of interest. Equivalent current noises $i_{nB}$, $i_{n1}$ and $i_{n2}$ will have the same power spectral density of $S_{ini} = 2qI_B$ [25].

Figure 5.6(b) illustrates a simplified equivalent model for the noise calculation that realistically assumes the drain-source conductance of each transistor is negligible compared to its $g_m$. First noise current $i_{nb}$ and $i_{n1}$ are combined and flow through the LP network comprising $C_1$ and $R_1$ ($=g_{m1}^{-1}$). The resulting noise voltage will be converted into output current noise by $g_{m1}$ and will combine with $-i_{n1}$ and $i_{n2}$ and together flow through another LP network, $C_2$ and $R_2$ ($= g_{s2}^{-1}$). Subsequently, output voltage noise $v_{no}$ appears at the output port. Note that the current noise of $M_1$ appears at both the input and the output ports of $g_{m1}$ ($i_{n1}$ and $-i_{n1}$, respectively), leading to two current sources that are fully correlated in the circuit model of Figure 5.6(b).

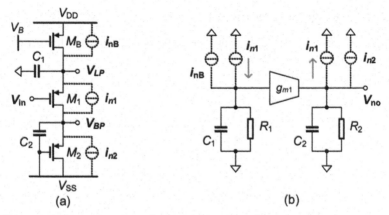

**Figure 5.6** BPF with noise sources. (a) Transistor circuit. (b) Equivalent model.

Following the aforementioned mechanism and approximating $g_{m1} \cong g_{s2} \cong g_{m2} = \frac{I_B}{n_p V_T}$, an average output noise power can be found from

$$\overline{v_{no}^2} = \int_0^\infty \left( S_{inB}|H_B(s)|^2 + S_{in2}|H_2(s)|^2 + S_{in1}|H_1(s)|^2 \right) df, \qquad (5.10)$$

where

$$H_B(s) = \frac{g_{m2}^{-1}}{D(s)} = \frac{g_{m2}^{-1}}{1 + s\left(\frac{C_1}{g_{m1}} + \frac{C_2}{g_{m2}}\right) + s^2\left(\frac{C_1 C_2}{g_{m1}g_{m2}}\right)}, \qquad (5.11)$$

$$H_2(s) = \frac{g_{m2}^{-1}}{1 + s\frac{C_2}{g_{m2}}}, \qquad (5.12)$$

and

$$H_1(s) = H_B(s) - H_2(s) = \frac{-s\frac{C_1}{g_{m1}g_{m2}}}{D(s)} \qquad (5.13)$$

After some mathematical rearrangement, this results in $\overline{v_{no}^2} = n_p kT/C_2$.

Note that (5.10–5.13) can be applied to the HP filter of Figure 5.3(a) by replacing $C_2$ by the gate-source parasitic capacitance of $M_2$ ($c_{gs2}$) since $c_{gs2}$ will bypass the output voltage to ground at very high frequencies eventually forming a BP response.

## 5.3 Cascaded Bandpass Filter

This section discusses the filter's topology selection, transistor-level circuit and relevant circuit characteristics.

### 5.3.1 Filter Topology Considerations

Figure 5.7(a) shows the $4^{\text{th}}$-order BPF topology used in [59] and [62]. It is composed of two identical $2^{\text{nd}}$-order sections connected in cascade. Capacitor $C_1$ with transconductor $G_{m1}$ and $C_2$ with $G_{m2}$ form the HP and LP cutoff frequencies of each $2^{\text{th}}$-order section, respectively.

To prevent loading by the input impedance of the subsequent stage, a voltage follower is inserted. This leads to more chip area and power consumption. In this work, we develop this structure further instead of the structures used in [60, 61] and [63], because this topology has a small number of active (and noisy) elements ($G_m$ cells) per noiseless elements ($C$). Therefore, lower noise and power consumption can be expected from this topology. In order to further enhance the filter's FoM, we need to eliminate the voltage buffer.

The proposed BPF biquad section in Figure 5.4 is compatible with the above requirement. Figure 5.7(b) shows the single-ended $4^{\text{th}}$-order BPF

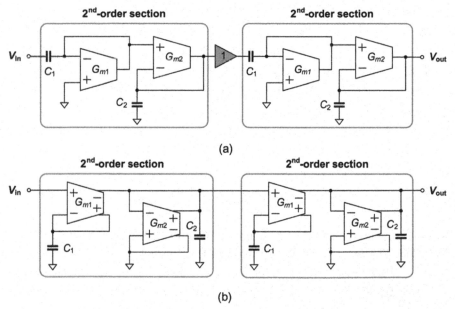

(a)

(b)

**Figure 5.7**  Cascaded $4^{\text{th}}$-order BPFs. (a) Topology of [59]. (b) Topology used in this work.

constituted by the two identical $G_m - C$ biquad sections proposed. The transfer function of each biquad section can be found by rearranging (5.6) to

$$H_{\mathrm{BP}}(s) = \frac{-s\frac{G_{m1}}{C_2}}{s^2 + s\left(\frac{G_{m2}}{C_2} + \frac{G_{m1}}{C_1}\right) + \frac{G_{m1}G_{m2}}{C_1C_2}}. \tag{5.14}$$

For $G_{m1} = G_{m2} = G_m$ we have,

$$\omega_o = \frac{G_m}{\sqrt{C_1C_2}}, \quad Q = \frac{\sqrt{C_1C_2}}{C_1 + C_2}, \quad K = \frac{C_1}{(C_2 + C_1)}, \tag{5.15}$$

where parameters $\omega_o$, $Q$ and $K$ stand for the center frequency, the quality factor and the mid-band gain, respectively. As this structure has a high-impedance input port, the cascade connection for higher order realization does not need an additional buffer circuit as long as the common-mode bias requirement is satisfied.

Note that this topology can only realize real poles and, as a consequence, its quality factor is limited to $Q = 0.5$. However, for a plethora of biomedical applications, e.g., in cochlear implant channels that require very low power consumption and electronic adjustability, this $Q$ is acceptable [59, 64].

## 5.3.2 Transistor Level Realization

At the transistor level, the filter topology in Figure 5.7(b) can be formed directly using the single-branch BPF of Figure 5.3(b) to implement the $2^{\mathrm{nd}}$-order sections. A center frequency that can be linearly adjusted by bias current $I_B$ is achieved. Transistors $M_1$ and $M_2$ are acting as $G_{m1}$ and $G_{m2}$, respectively. Note that the equivalent operation of the proposed filter topology and the circuit is valid only for the small-signal condition. As a consequence, a differential structure as shown in Figure 5.8 is required to maximize the filter's dynamic range. The power consumption and output average noise power defined in (5.10–5.13) will double here as well as the voltage signal swing for the differential version in Figure 5.8. In this case bias current $I_B$ defines the transconductance

$$G_{m1} = \frac{g_{m1}}{2} = G_{m2} \cong \frac{g_{m2}}{2} = \frac{I_B}{2n_pV_T}. \tag{5.16}$$

In line with (5.15), $\omega_o$ is linearly adjustable via $I_B$.

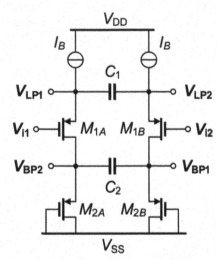

**Figure 5.8**   Differential version of the single-branch BPF.

### 5.3.3 Dynamic Range

We estimate the DR of the BPF shown in Figure 5.7(b) from the harmonic components of the filter induced by the nonlinear (exponential) behavior of input transistors $M_{1A}$ and $M_{1B}$ when an input signal with frequency ($f_{in}$) equal to $f_C$ of the filter is applied. Operating at this frequency, the impedance of $C_1$ is close to zero. This causes the source terminals of $M_{1A}$ and $M_{1B}$ to be connected together. For this reason, $V_{GS1A}$ and $V_{GS1B}$ will experience the largest input signal amplitude compared to the case in which an input voltage with a lower frequency but with the same amplitude is applied. For higher frequencies, similar harmonic components will be generated by $M_{1A}$ and $M_{1B}$, however, they will be filtered out by $M_{2A}$ and $M_{2B}$ and $C_2$. As a consequence, the harmonic components that appear at the output will be the highest when $f_{in} = f_C$.

Now $M_{1A}$ and $M_{1B}$ can be seen as a source-coupled pair transconductor biased by a current source that equals $2I_B$. Assuming $M_{1A}$ and $M_{1B}$ are matched perfectly and operated in weak inversion saturation, we can find that

$$I_{od} = I_{D1B} - I_{D1A} = 2I_B \tanh\left(\frac{V_{id}}{n_p V_T}\right). \qquad (5.17)$$

This differential current is injected into an active load, comprising $M_{2A}$ and $M_{2B}$ and capacitor $C_2$, and a differential output voltage $V_{od}$ will be produced. Hence, the distortion of $V_{od}$ is caused by the nonlinear behaviors of both $I_{od}$ and the active load. By setting $C_1 > 10C_2$ to achieve a $K$ close to $g_m g_s^{-1} = n_p^{-1}$, the voltage swing at the output is always less than the voltage swing at the input. In order to estimate the linear input range of the filter, we assume further that this output voltage swing is sufficiently small for $M_{2A}$ and $M_{2B}$ to produce negligible harmonic components, and that $I_{od}$ can be considered the dominant source of distortion.

Applying a Taylor series expansion to (5.17) and considering only the first two terms, we have

$$\frac{I_{od}}{2I_B} = \frac{V_{id}}{n_p V_T} - \frac{1}{12}\left(\frac{V_{id}}{n_p V_T}\right)^3. \tag{5.18}$$

In the case that a sinusoidal $V_{id}$ with amplitude $V_p$ is applied to (5.18), the $3^{rd}$-harmonic distortion can be calculated to be

$$HD_3 = \frac{1}{48}\left(\frac{V_p}{n_p V_T}\right)^2. \tag{5.19}$$

For $HD_3$ of 1%, $V_P = 0.69 n_p V_T$ (i.e., $1.1 V_T$ for $n_p = 1.6$). Hence, we can find the maximum input signal power of a sinusoidal output signal with $f_{in} = f_C$ as

$$P_{sig} = K^2 \frac{V_p^2}{2} = \frac{V_p^2}{2n_p^2} = 0.24 V_T^2. \tag{5.20}$$

Referring to Section 5.2.3, the output noise power for the circuit in Figure 5.8 can be found to be $v_{noDiff}^2 = 2n_p kT/C_2$. Then we can approximate the DR of the BPF circuit shown in Figure 5.8 as

$$DR_{second} \cong 10\log\left(\frac{0.24 C_2 V_T^2}{2n_p kT}\right) \cong 10\log\left(\frac{0.075 C_2 V_T^2}{kT}\right), \text{ for } n_p = 1.6. \tag{5.21}$$

For the $4^{th}$-order BPF obtained by cascading two identical $2^{nd}$-order sections of Figure 5.8 with $K_i$ for each stage, the output noise power of the entire filter circuit can be found from

$$\overline{v_{noFourth}^2} \simeq \overline{v_{noDiff}^2} + \overline{v_{noDiff}^2} K_2^2 \cong \overline{v_{noDiff}^2}\left(1 + K^2\right). \tag{5.22}$$

Thus, the DR of the $4^{th}$-order BPF can be calculated to be:

$$\text{DR}_{\text{Fourth}} \cong 10\log\left(\frac{0.12C_2V_p^2}{n_pkT\left(1+n_p^2\right)}\right) \cong 10\log\left(\frac{0.054C_2V_p^2}{kT}\right)$$

for $n_p = 1.6$. \hfill (5.23)

### 5.3.4 Common-Mode Behavior

It is interesting to analyze the common-mode behavior of the differential circuit that provides two filtering functions in Figure 5.8. For the LPF, the output voltages are taken from the source terminals of the circuit configured as source followers. In the pass-band the output voltages will follow the input voltages regardless whether differential-mode or common-mode signals are applied. There is no common-mode rejection at all.

On the other hand, the output voltages of the BPF are taken from the drain terminals of $M_1$ and as a result, assuming $M_{1A}$ and $M_{1B}$ are matched perfectly, high common-mode rejection capability can be expected. Theoretically, under the condition that each current source $I_B$ has infinite output impedance and neglecting the channel length modulation effect of $M_1$, the output common-mode and differential mode signal will be completely isolated thus featuring an infinite common-mode rejection ratio.

In practice, a low-frequency common-mode gain of this circuit can be found for the case in which $M_1$ and $M_2$ are perfectly matched, which equals $A_{\text{CM}} \cong -g_{oc}g_{m2}^{-1}$, where $g_{oc}$ represents the output conductance of current source $I_B$. Enhancing the common-mode rejection ratio (CMRR) can be done by improving the output impedance of $I_B$. Besides, due to mismatch between transistor pairs $M_1$ and $M_2$, common-mode to differential-mode conversion will occur and the CMRR will be further degraded.

## 5.4 Design Methodology

This section presents the design procedure for a BPF for application in an analog bionic ear processor. In this particular application, the following issues should be considered: filter order, midband gain, DR and $f_C$.

### 5.4.1 Filter Order

For the practical use of a BPF in a bionic ear processor as was suggested in [59], a transition band rolloff of 40 dB per decade is sufficient. For this reason, the filter topology shown in Figure 5.7(b) can be employed and its $2^{\text{nd}}$-order section replaced by the BPF circuit shown in Figure 5.8. Thus, a $4^{\text{th}}$-order BPF with a total current consumption of $4I_B$ is formed.

## 5.4.2 Midband Gain and Dynamic Range

For practical use of this BPF in a bionic ear, a logarithmic compressor circuit block will be placed in front of the BPF. This is to imitate the biological operation of the cochlear outer hair cell that can handle a large dynamic range (more than 100 dB) of the incoming sound pressure [65]. For this reason, a BPF filter with moderate dynamic range of around 40 dB should be sufficient to handle the output signal of the logarithmic compressor.

To obtain a DR of 40 dB, a $C_2$ of approximately 0.5 pF is calculated from (5.23). However, the value of $C_2$ should be chosen higher than the calculated value to compensate for the error from this approximation. Hence, a $C_2$ of 0.6 pF is used. To make the DR approximation valid, $K$ needs to be maximized. Considering the midband gain of the $2^{nd}$-order section in (5.15), for a cascaded BPF structure used here, we found that $K = \left(\frac{C_1}{C_1+C_2}\right)^2$. We thus set $C_1 = 11C_2 = 6.6$ pF. With this setting a quality factor $Q$ of 0.28 is obtained using (5.15). This is a fundamental tradeoff between midband gain and quality factor for this BPF.

## 5.4.3 Center Frequency, Bias Current and Tuning

In a cochlear speech processor, the target frequency range for the BPF is from 100 Hz to 10 kHz. As the capacitances have been specified already above, $f_C$ is now controlled by $I_B$. It can be found from (5.15) and (5.16) that

$$f_C = \frac{I_B}{4\pi n_p V_T \sqrt{C_1 C_2}}. \tag{5.24}$$

For $f_C = 1$ kHz, we obtain $I_B \cong 1$ nA from (5.24) and $I_B$ can be adjusted linearly to obtain other values of $f_C$ (i.e, 0.1 nA and 10 nA for $f_C$ of 100 Hz and 10 kHz, respectively).

Due to process variations, around 30% $f_C$ variation might be obtained after fabrication. 10% of these variations can originate from integrated capacitors and 20% from transconductance values of the transistor [66]. For application in a bionic ear BPF that does not require a very precise value for its center frequency. We suggest that 10% variation of $f_C$ should be acceptable. Hence, a tuning loop that utilizes a fixed external resistor, $R_{ext}$, to calibrate the value of bias current $I_B$, thereby regulating the variation of $f_C$ within 10% (resulting from the integrated capacitor only) is recommended. Figure 5.9(a) shows the tuning loop that can be used. A precise $R_{ext}$ is connected off-chip. $g_{mT}$ is an extra transconductor having the same transconductance as the transconductor used in the BPF circuit. If the current flowing through $R_{\text{ext}}$ is not identical to the output current of $g_{mT}$, the resulting error current will be integrated by capacitor $C$ resulting in voltage $V_{\text{ctr}}$. $V_{\text{ctr}}$ will adjust the on-chip bias circuit to generate bias current $I_{\text{Bias}}$ to update the value of $g_{mT}$. After settling, the relationship $g_{mT} = R_{\text{ext}}^{-1}$ will be obtained. At the same time, $I_B$, which is the bias current of the BPF will be updated as well. Eventually the transconductance of the transistors used in the BPF circuit will be controlled by the more reliable external resistor $R_{\text{ext}}$ in a master-slave fashion and an $f_C$ variation of approximately 10% can be obtained. Transconductor $g_{mT}$ can be realized by the circuit in Figure 5.9(b). If $M_{1A}$ and $M_{1B}$ are matched and identical to the pMOS transistor pair in Figure 5.8, this circuit yields sufficient performance.

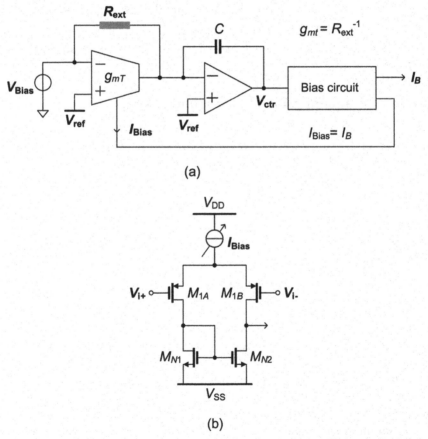

**Figure 5.9** Possible $f_c$ tuning. (a) Tunning loop and (b) controlled $g_{mt}$ circuit.

### 5.4.4 Transistor Dimensions

To make all of the design assumptions valid, all transistors used should be sized such as to have noise corner frequency (where the flicker noise and shot noise have the same spectral density) below $0.1 f_C$. Since the dimensions of the transistors are (theoretically) orthogonal to other filter specifications, we can set them unreasonably large at first and then reduce them, while still meeting all specifications in circuit simulation. By doing so, we obtain a transistor width ($W$) and length ($L$) of 10 $\mu$m and 2.5 $\mu$m, respectively, for $M_{1A}$, $M_{1B}$, $M_{2A}$ and $M_{2B}$. For transistors that form current source $I_B$, we set $W = 24$ $\mu$m and $L = 6$ $\mu$m. This is also to have a good matching among the transistors used in the bias circuit.

## 5.5 Measurement Results

The proposed $4^{\text{th}}$-order BPF shown in Figure 5.7(b) has been fabricated in 0.18-$\mu$m AMS CMOS technology with a nominal threshold voltage of $V_{tp} \cong -0.42$V. The filter chip photo is shown in Figure 5.10. Including the filter

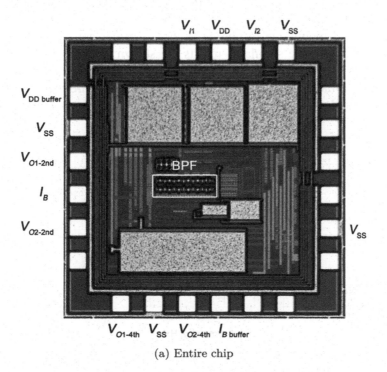

(a) Entire chip

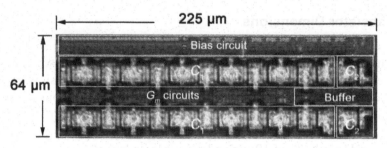

(b) Zoomed-in area of the BPF

**Figure 5.10**  Chip photo.

core (pMOS transistors and MIM capacitors $C_1 = 11C_2 = 6.6$ pF), the bias
circuit (formed by simple current mirror circuits) and source follower buffers
to drive the off chip capacitive load (formed by the pads, chip package, PCB
and instrument probes), the chip occupies 64 $\mu$m × 225 $\mu$m. The following
results were measured using a dynamic signal analyzer (SR785) under the
condition of $V_{DD} = 0.5$ V for the BPF and bias circuits, $V_{DD} = 1.8$ V for the
buffers and $V_{CM}$ set to 0.15 V. An external bias current is supplied from a
precision current source (Keithley 6430).

The measured magnitude responses of the $2^{nd}$-order and the $4^{th}$-order
BPFs compared with the simulated results are shown in Figure 5.11. Bias
current $I_B$ was set to 1 nA to obtain a 1 kHz $f_C$. From the measurement
(black line), mid-band gains $K$ of –1.54 dB and –2.63 dB are observed for
the $2^{nd}$-order and $4^{th}$-order filters, respectively. These values include the

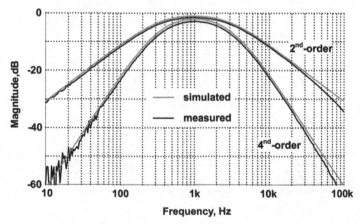

**Figure 5.11**  Measured and simulated magnitude responses of the proposed BPFs.

loss of the source follower buffers, estimated to be around –0.45 dB by comparing with the simulations without buffer (gray line).

Figure 5.12 shows the measured (black line) and the simulated (gray line) magnitude responses of the $4^{th}$-order BPF for $I_B$ ranging from 0.125 nA to 16 nA. $f_C$ moves almost linearly for 7 octaves, starting from 124 Hz to 15.8 kHz. We can observe from the measured results that, for $I_B$ lower than 0.5 nA, $K$ starts decreasing. At these values of bias current, diode-connected $M_2$ is forced to leave the weak inversion saturation region as its drain-source voltage is being reduced by $I_B$. It affects the magnitude response and gives a lower limit to the filter's adjustability. The upper limit is defined by $V_{DD}$. When $I_B$ goes high, the gate-source voltages of all transistors will go up, and after it reaches a certain value, the source voltage of $M_1$ and $V_{DD}$ will start forcing $M_B$ out of its saturation region. This implies that the tuning range for higher frequency can be widened by supplying more $V_{DD}$. Unfortunately, these behaviors cannot be predicted precisely by simulations. Also noise from the measurement setup can be noticed at a magnitude of around –60 dB at frequencies lower than 200 Hz. This is because the signal amplitude was set very small (10 mV) within the linear range of the BPF. Low frequency noise with peak values around 10 $\mu$V affect the measured results in this range.

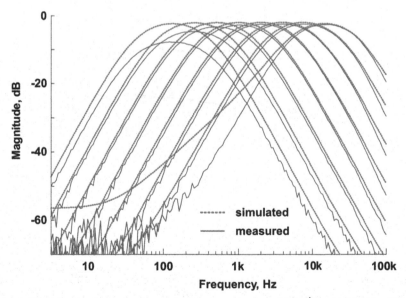

**Figure 5.12**  Measured and simulated magnitude responses of the $4^{th}$-order BPF for different bias current ranging from 0.125 nA to 16 nA.

To see the tunability of the $4^{\text{th}}$-order BPF in more detail, center frequencies obtained for different values of $I_B$ have been collected and plotted in Figure 5.13. Linear tunability of the proposed filter is obtained for 7 octaves (more than 2 decades). As has been discussed in the previous paragraph, although the linear adjustability is confirmed for $f_C$, mid-band gain $K$ cannot be maintained constant over the whole tuning range. This phenomenon can be seen more clearly from Figure 5.14, in which values of $K$ have been collected and plotted for the same conditions as for the results obtained in Figure 5.14.

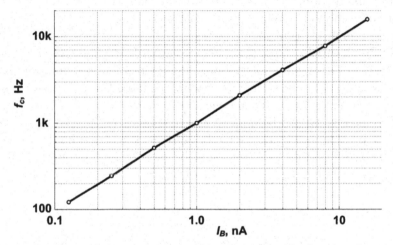

**Figure 5.13**    Measure center frequency versus bias current.

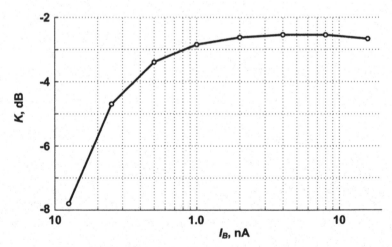

**Figure 5.14**    Measured midband gain versus bias current.

For $I_B$ less than 1 nA, $K$ drops below −3 dB. On the other hand, for higher $I_B$, a $K$ higher than −3 dB can be maintained.

Figure 5.15 shows the measured and simulated output noise voltage spectral densities of both BPFs for $I_B = 1$ nA and $f_C = 1$ kHz. At frequencies lower than $f_C$, both shot noise and flicker noise contribute to the output. This flicker noise is from the source follower buffers that are biased by a DC current of 10 $\mu$A to drive offchip loads so the noise corner frequency of the source follower transistors is higher than that off the transistors in the BPF. For frequencies higher than 1 kHz, only shot noise plays a role and it is suppressed by the filter's transfer function. Integrated over the entire bandwidth, measured output noise voltages of 54 $\mu V_{rms}$ and 57.4 $\mu V_{rms}$ are obtained for the $2^{nd}$-order and $4^{th}$-order BPFs, respectively. Approximately 3 dB difference in the shot noise can be observed from the measurement and simulation.

Measured and simulated output noise voltage spectral densities of the $4^{th}$-order BPF for different $f_C$ (adjusted by changing $I_B$) are presented in Figure 5.16. The filter noise density goes lower for higher $I_B$ (higher $f_C$ and bandwidth). This mechanism maintains the same integrated shot noise power for different values of $I_B$.

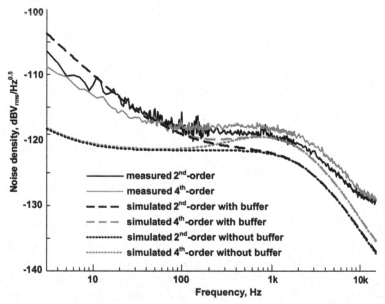

**Figure 5.15** Measured and simulated output noise voltage spectral density for $I_B = 1$ nA and $f_c = 1$ kHz.

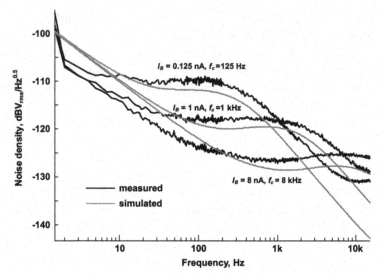

**Figure 5.16**  Measured and simulated output noise voltage spectral density for different center frequencies.

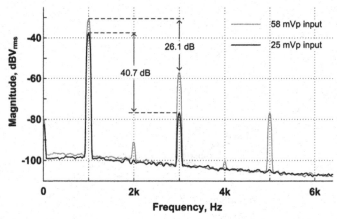

**Figure 5.17**  Output voltage spectra of the $4^{\text{th}}$-order BPF for $f_{\text{in}} = f_c = 1$ kHz.

The linearity of the filter has been tested by applying a sinusoidal input voltage to the filter with input frequency $f_{\text{in}} = f_C$ and observing its output spectrum. For the case of $I_B = 1$ nA (1 kHz $f_C$), the measured results for input amplitudes ($V_{\text{inp}}$) of 25 mV and 58 mV are illustrated in Figure 5.17. Since the proposed filter operates in a differential fashion, the $3^{\text{rd}}$ harmonic component was found to be the main harmonic component. The $3^{\text{rd}}$ harmonic distortions (HD$_3$) were found at –40.7 dB and –26.1 dB for 25 mV and 58 mV $V_{inp}$s, respectively. These values are associated with a total harmonic distortion (THD) of 1% and 5%, respectively.

Figure 5.18 provides the values of second harmonic distortion HD$_2$ and HD$_3$ of both the 2$^{nd}$-order and the 4$^{th}$-order filters for different $V_{inp}$ at $f_{in}$ = $f_C$ = 1 kHz. For the range of 20 mV < $V_{inp}$ < 60 mV, HD$_2$ appeared more than 20 dB below HD$_3$ for both cases. For this reason HD$_3$ can be considered to be responsible for the THD. The HD$_3$ for the 4$^{th}$-order BPF was found 3 dB worse than that of the 2$^{nd}$-order BPF for the entire range of $V_{inp}$. As the filter's pass band is quite flat and the transition band roll-off is not sharp, the third order intermodulation distortion (IMD$_3$) can be estimated by calculating it from the exponential behavior of transistor pair $M_1$. For $V_{inp}$ = 25 mV and $V_T$ = 26 mV, it is calculated to be IMD$_3$ $\cong$ –28.5 dBc. This value is inline with the value of HD$_3$ found in 5.17 (IMD$_3$ $\cong$ 10 dB+HD$_3$).

Other relevant filter parameters at $I_B$ = 1 nA were also tested and are summarized for the 2$^{nd}$ and 4$^{th}$-order BPFs in Table 5.1. It can be seen that all the filter characteristics of the 2$^{nd}$-order BPFs are better than those of the 4$^{th}$-order one except the filter selectivity. This is due to the numbers of transistors and bias current branches in the 2$^{nd}$-order BPF are less than in the 4$^{th}$-order BPF. More detail on the 4$^{th}$-order BPF characteristics are summarized in Table 5.2 for three different cutoff frequencies.

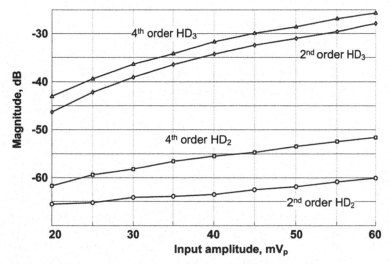

**Figure 5.18** Harmonic components versus input amplitude for 1 kHz $f_{in}$ and $I_B$ = 1 nA ($f_c$ = 1 kHz).

**Table 5.1**   Measured filter performance for 1 kHz center frequency

| Filter | 2nd-Order (Measured) | 4th-Order (Measured) | 2nd-Order (Simulated) | 4th-Order (Simulated) |
|---|---|---|---|---|
| Total current [nA] | 2 ($2I_B$) | 4 ($2I_B$) | 2 ($2I_B$) | 4 ($2I_B$) |
| $V_{DD}$ [V] ; $P$ [nW] | 0.5 ; 1 | 0.5 ; 2 | 0.5 ; 1 | 0.5 ; 2 |
| $K$ [dB] | −1.54 | −2.63 | −1.03 | −2.05 |
| BW ($f_l$–$f_h$) [Hz] | 290–3.49 k | 420–2.55 k | 290–3.5 k | 420–2.55 k |
| $V_{inp}$ @1% ; 5% THD [mV] | 29 ; 68 | 25 ; 58 | 27 ; 65 | 26 ; 58 |
| *Output noise [$\mu V_{rms}$] | 54 | 57 | 37 | 40 |
| DR @ 1% ; 5% THD [dB] | 50 ; 58 | 47 ; 55 | 54 ; 62 | 53 ; 60 |

*Integrated over BW

**Table 5.2**   Measured performance summary of the proposed filter

| $I_B$ [nA] | 0.125 | 1 | 8 |
|---|---|---|---|
| $f_C$ [Hz] | 124 | 1 k | 7.84 k |
| BW ($f_l$ – $f_h$) [Hz] | 48–345 | 420–2.55 k | 3.15 k–19.7 k |
| $K$ [dB] | −7.85 | −2.63 | −2.55 |
| *Output noise [$\mu V_{rms}$] | 51 | 57.4 | 67 |
| Inp. referred noise [$\mu V_{rms}$] | 126 | 78 | 90 |
| $V_{inp}$ @1% ; 5% THD [mV] | 27 ; 67 | 25 ; 58 | 25 ; 59 |
| DR @1% ; 5% THD [dB] | 44 ; 52 | 47 ; 55 | 46 ; 53 |
| FoM @1% ; 5% THD [$10^{-18}$J] | 20 ; 3.2 | 10 ; 1.6 | 12.8 ; 2.6 |

*Integrated over BW.

Table 5.3 shows a performance comparison among existing biomedical BPFs collected from journal articles with measurement results. The main distinct features of this design are the 0.5 V $V_{DD}$, smallest chip area and FoMs of $9.98 \times 10^{-18}$ J and $1.58 \times 10^{-18}$ J measured at 1% THD and 5% THD, respectively. The latter value is approximately an order of magnitude better than that of [62], the lowest number reported until recently. Also our BPF occupies approximately 10 times smaller chip area compared with that of [62]. The BPF of [62] and ours are comparable in circuit complexity and process technology but the pMOS 2$^{nd}$-order circuit cell of [62] cannot be connected in cascade without considerable loading effect and the circuit itself requires a higher $V_{DD}$ of $2V_{SG} + V_{SDsat}$. It is also interesting to see that the 7$^{th}$-order BPF of [63] consumes extremely little power of 60 pW which is almost 45 times smaller than that of our design, but it does not provide the best FoM since its $f_C$ is only 2 Hz. It is worth mentioning that the FoM used may be too simplistic when the other parameters of the filters, i.e, sensitivity, type of magnitude response, application, etc., are considered. As can be seen

**Table 5.3** Performance summary and comparison

| Reference | [59], 2003 | [60], 2007 | [61], 2009 | [62], 2010 | [63], 2011 | This Work |
|---|---|---|---|---|---|---|
| CMOS tech. [$\mu$m] | 1.5 | 0.35 | 0.18 | 0.18 | 0.35 | 0.18 |
| Order | 4 | 6 | 8 | 4 | 7 | 4 |
| Chip area [mm$^2$] | NA | 0.234 | 0.11 | 0.132 | 0.216 | 0.0144 |
| $f_C$ [Hz] | 141 | 671 | 3.5 k | 732 | 2 | 1 k |
| $P$ [nW] | 230 | 68 | 875 | 14.4 | 0.06 | 2 |
| $V_{DD}$ [V] | 2.8 | 1 | 1.2 | 1 | 1 | 0.5 |
| Inp. Referred noise [$\mu V_{rms}$] | 776 | 50 | NA | 50 | 51 | 78 |
| THD [%] | 5 | NA | NA | 1 | 0.3 | 1  5 |
| DR [dB] | 67.5 | 49 | 37 | 55 | 43 | 47  55 |
| FoM [$10^{-18}$ J] | 72.5 | 213 | 6240 | 15.6 | 215 | 9.98  1.58 |
| Dedicated application | Cochlear implant | Breathing detector | Gabor transform | Biopotential recorder | Wavelet transform | Cochlear implant |

from [63] that presents a sophisticated wavelet filter consuming almost zero power, FoM obtained is still worse than this design which only implements a simple $4^{th}$-order transfer function. Therefore, a better figure of merit should be developed to make a more reasonable comparison.

## 5.6 Conclusions

A smart choice of the filter topology and a very compact circuit that operates from a very low supply voltage are the keys to the design of a BPF that achieves a good FoM. Measurement results of the proposed BPF filter, designed according to the keys mentioned above, show a considerable FoM improvement with respect to other existing designs. The proposed BPF filter can find its application in multi-channel cochlear implant speech processors that require very low power consumption and more than 6 octaves tuning ability ranging from 100 Hz to 10 kHz [59]. Although the DR of this proposed BPF is not as high as that of the BPF in [59], in combination with a logarithmic compressor as recently suggested in [65], a sufficient overall DR can be obtained.

# 6

# Follower-Integrator-based
# LPF for ECG Detection

"The most complicated skill is to be simple."
— Dejan Stojanovic

## 6.1 Introduction

In the design of fully-integrated $Gm - C$ biomedical filters, realizing very large time constants is one of the major challenges. Using conventional strong inversion CMOS devices, either capacitance multiplier [67] or transconductance reduction [68] techniques need to be applied to obtain a cutoff frequency of less than a few hundred hertz. Due to the device's operating region that conducts current in the range of $\mu$A and the techniques used, which give rise to the filter's circuit complexity, power consumptions in the range of few $\mu$W are required for these kinds of filters.

To reduce power consumption, subthreshold CMOSTs that conduct less current can be used without the aforementioned techniques. Since the transconductance of a single transconductor in subthresholds is sufficiently low, forming large time constants with on chip capacitors becomes possible [69]. Unfortunately, however, nonlinearity, noise and mismatch in weak inversion are more severe compared to those aspects in strong inversion [40]. Recently, the systematic design of a LPF for portable ECG applications has been introduced [70]. The filter provides a good figure-of-merit (FoM) and consumes a very low power of 453 nW. However, the filter in [70] relies on a linearized $G_m$ that comprises several transistors and suffers from a high amount of noise and mismatch-induced nonlinearity. To maintain the filter's dynamic range (DR) in the noisy circuit structure, the power consumption cannot be reduced further.

This chapter describes a LPF that does not require transconductor linearization. The filter thus benefits from fewer noise sources and fewer current

83

consuming branches. Furthermore, a nonlinear $g_m$ with negative feedback is used to create a $1^{st}$-order LPF before forming the higher-order filter by a cascade connection of 6 identical stages. Due to this local feedback topology, good linearity is expected in the filter's passband. For this reason, the desired DR of the filter can be obtained with lower power consumption compared to [70]. Note that $g_m$ is used here to indicate the small-signal transconductance of the nonlinear transconductor, which is different from the transconductance $G_m$ of the linearized transconductor that can handle a wider linear range.

At circuit level, a follower integrator (FI) [70], which is a special class of the single-branch filters presented in Chapter 5, is employed as the $1^{st}$-order section [9]. By doing so, linear adjustability of the cutoff frequency can be obtained using its bias current. The filter has been fabricated in AMS 0.18-$\mu$m CMOS technology. The filter chip has been measured while operating from supply voltages ($V_{DD}$) of 0.5 V, 0.6 V and 0.7 V. Filter characteristics including power consumption ($P$), order ($N$), DR, and bandwidth ($f_c$) have been collected to estimate the filter's FoMs according to the following formula:

$$\text{FoM} = \frac{P}{N \cdot f_c \cdot \text{DR}}.$$ (6.1)

Figure 6.1 shows the FoMs obtained for three different values of $V_{DD}$ compared to other existing biomedical LPFs. For all cases, the proposed filter provides almost three orders of magnitude improvement with respect to a state-of-the-art design [70].

In the next section, background information required to understand the design of a LPF as required in an ECG recording system will be presented. Section 6.3 shows the basic concept of the FI circuit along with a discussion of its transistor level realizations. In Section 6.4, the $6^{th}$-order LPF design will be described together with its performance analysis and design methodology. Measurement results are discussed and compared to other existing designs in Section 6.6. Finally, Section 6.7 summarizes the work on FI-based LPFs .

## 6.2 ECG Detector LPF Design

Figure 6.2 shows a general block diagram of a portable ECG detection system. It comprises a low noise amplifier, a continuous-time LPF (this design), an

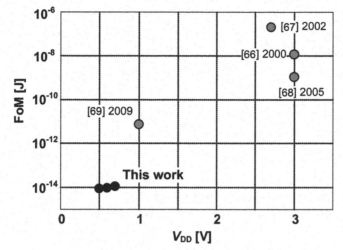

**Figure 6.1** FoM comparison of biomedical LPFs.

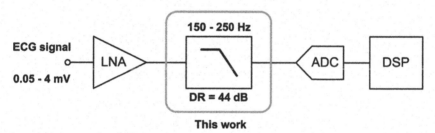

**Figure 6.2** Portable ECG detection concept.

analog to digital converter (ADC) and a digital signal processor (DSP). The low-noise amplifier is used to amplify the very weak ECG signal. Depending on the electrodes used, the ECG signal amplitudes can range from 50 $\mu$V to approximately 4 mV (see Figure 6.3) [72]. Next, high frequency components in the ECG signal are filtered out to decrease the out-of-band noise. The recommended lowpass cutoff frequencies are 150 Hz and 250 Hz for adults and children, respectively [73, 74]. Afterwards, an ADC is utilized in which the analog input signal is quantized and converted into digital values as needed for the subsequent DSP.

Since there is a large variation in the expected amplitude of the input signal, the amplifier and filter are required to have a minimal DR, according to [70] and [75], of

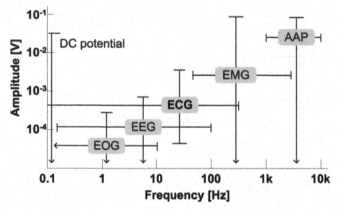

**Figure 6.3**   Different bio-potential voltages versus frequency spectrum. Adapted from [72].

$$DR = 20 \log \left( \frac{2ECG_{max}}{ECG_{min}} \right) = 20 \log \left( \frac{2 \times 4\,mV}{50\,\mu V} \right) = 44\,dB. \qquad (6.2)$$

To avoid aliasing, the LPF should provide as much attenuation in the stop-band as possible. Moreover, a constant group delay over the passband response of the filter should be obtained to minimize phase distortion phase distortion [76].

## 6.3 Follower-Integrator-based Lowpass Filter

This section provides a fundamental concept of the FI and discusses the FI circuit realization at transistor level.

### 6.3.1 Concept

The FI is shown in Figure 6.4(a). It comprises a transconductor and a grounded capacitor connected in a negative-feedback fashion. Resistor $R_o$ represents the output resistance of the transconductor to ground (which is usually many times larger than $g_m^{-1}$). We can also represent the circuit in the form of the feedback block diagram shown in Figure 6.4(b), where $Z_o$ represents a parallel connection of $C$ and $R_o$. It can be seen that the circuit's loop gain (LG) equals $G_m Z_o$.

Assuming $g_m R_o \gg 1$, the transfer function of this circuit is given by

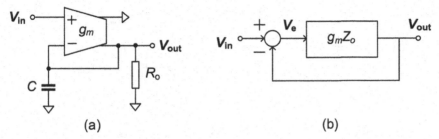

**Figure 6.4** Follower integrator. (a) Macro-model and (b) feedback block diagram.

$$H_1(s) = \frac{V_{\text{out}}(s)}{V_{\text{in}}(s)} = \frac{LG}{1+LG} = \frac{g_m Z_o}{1 + g_m Z_o} = \frac{\frac{g_m R_o}{1 + g_m R_o}}{1 + \frac{sCR_o}{1+g_m R_o}}$$

$$\cong \frac{1}{1 + s\frac{C}{g_m}}, \text{ for } g_m R_o \gg 1. \qquad (6.3)$$

As can be seen from (6.3), the FI provides a lowpass frequency response with a passband gain and cutoff frequency of $K \cong 1$ and $f_c = g_m/2\pi C$, respectively. According to this characteristic, $V_{\text{out}}$ is following $V_{\text{in}}$ closely for input signal frequencies below $f_c$. In other words, the differential input voltage of the $g_m$ block is kept small, which helps the filter to suffer less from the nonlinearity of the $g_m$ itself. It should be noted here that benefitting from the fact mentioned above, high DR LPFs realized from transconductors without linearization have been successfully implemented in [77] and [35]. Hence, without linearization, the $G_m$ can be made compact and low noise and low-power consumption can be expected, while a good linearity can be achieved due to the large loop gain at low frequencies.

## 6.3.2 Transistor-Level Consideration

The most compact solution to realize the FI is shown in Figure 6.5(a). The concept of single branch filters in Chapter 5 is extended here. Source follower (SF) $M_1$ and its bias current $0.5I_B$ form the negative feedback $G_m$. Requiring only one current source to supply its single branch makes this circuit suitable for very low power design. However, the voltage swing at the circuit input is limited to a few mVs by the nonlinearity of $M_1$. To enlarge the signal swing, a differential version is required, as shown in Figuer 6.5(b). Compared with the single-branch SF topology, the differential version quadruples the power consumption to achieve the same cutoff frequency from the same value of

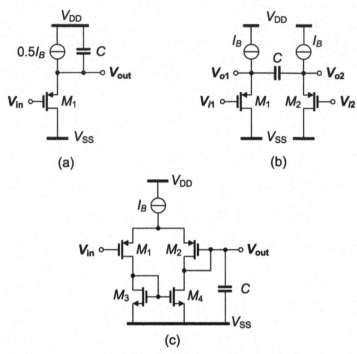

**Figure 6.5**  FI circuits using: (a) single-branch SF, (b) differential SF and (c) source-coupled pair (SCP) transconductor.

capacitor $C$. This is due to the fact that the transconductances of $M_1$ and $M_2$ (operating in weak inversion saturation) are linearly proportional to their drain currents, and to achieve the same effective transconductance as the single-branch SF, the bias currents of $M_1$ and $M_2$ in Figure 6.5(b) must be changed from $0.5I_B$ to $I_B$. As a consequence of this bias point setting, the input-referred average noise powers of the single-ended and the differential versions are identical.

For the high-order filter application presented in this chapter, a cascade connection of circuit cells in Figures 6.5(a) and 6.5(b) is needed. However, (as already discussed in the previous chapter) since the common-mode levels of input and output signals are different, cascading these cells directly cannot be made without the requirement for either a higher supply voltage ($V_{DD}$) or using complementary devices that have different transconductances for the same drain current. As a consequence, mismatch between the slope factors of nMOS and pMOS devices will introduce an error in the filter's frequency response.

To allow for the cascaded connection of the FIs to realize a high-order filter, the differential pair transconductor connected in a negative feedback fashion with capacitive load $C$ shown in Figure 6.5(c) can be employed [71]. Note that the negative feedback also exists in the SF topologies in Figures 6.5(a) and (b) but it is not explicit. The negative feedback mechanism happens at the source terminal of a single transistor. For the circuit in Figure 6.5(c), transistors $M_1$–$M_4$ with bias current $I_B$ form an open-loop transconductor. The explicit connection between the drain and gate terminals of $M_2$ provides negative feedback.

As the even-order harmonic components do not exist in the large-signal characteristic of the source-coupled pair (SCP) transconductor operating in weak inversion saturation (defined by the hyperbolic tangent function after neglecting all non-idealities of the current mirror active load), this circuit is equivalent to the differential SF lowpass filter in Figure 6.5(b) in terms of linearity. To achieve the same $f_c$ as the SF LPF in Figure 6.5(a) from the same value of $C$, this circuit doubles current consumption and quadruples noise power. Obviously, the SCP-based FI outperforms the circuit shown in FIgure 6.5(b) in terms of power consumption, but the output noise power is quadrupled. For this design, a cascade connection operating from a very low supply voltage is required and therefore the FI circuit in Figure 6.5(c) is chosen.

Let's examine the FI circuit Figure 6.5(c) in more detail. It comprises a differential pair $M_1$–$M_2$ biased by current source $I_B$ with a current mirror active load, $M_3 - M_4$, and a grounded capacitor, $C$. Neglecting channel length modulation, the current flowing through capacitor $C$ can be found to be

$$I_C = I_B \tanh\left(\frac{V_{\text{in}} - V_{\text{out}}}{2n_p V_T}\right). \tag{6.4}$$

At low frequencies, $I_C \cong 0$ and consequently, $V_{\text{in}}$ and $V_{\text{out}}$ are forced to be equal by the unity gain negative feedback mechanism. The hyperbolic tangent function will not produce any distortion in this case. For higher frequencies, the phase difference between $V_{\text{in}}$ and $V_{\text{out}}$ becomes greater and more distortion will be produced.

Assuming the slope factors of nMOS and pMOS are identical, i.e., $n_p = n_n = n$, at low frequency, the power spectral density of the output noise voltage can be found as

$$S_{vno}(f) = \frac{8nkT}{g_m} + \frac{2}{f}\left(\frac{K_{Fn}}{WL_1} + \frac{K_{Fp}}{WL_3}\right), \tag{6.5}$$

where $K_{Fn}$ and $K_{Fp}$ are the flicker noise parameters of the nMOS and pMOS devices, respectively. All the other symbols have their usual meanings, $WL_1 = WL_2$, $WL_3 = WL_4$ and $g_m \cong I_B/2nV_T$.

For ECG applications that deal with input frequencies below a few hundred hertz, the filter's noise power is not only contributed by the shot noise in weak inversion but also by $1/f$ noise. From (6.5), the noise corner frequency where the power spectral density of the $1/f$ noise and the shot noise are equal is given by

$$f_{\text{corner}} = \frac{qI_B}{8(nkT)^2} \left( \frac{K_{Fn}}{WL_1} + \frac{K_{Fp}}{WL_3} \right). \tag{6.6}$$

We can see from (6.6) that the FI requires large area transistors and a low bias current to keep $f_{\text{corner}}$ low.

## 6.4 ECG Lowpass Filter Design

This section discusses the ECG filter topology, analyzes the filter performance and presents the filter design methodology.

### 6.4.1 Filter Topology

The requirements of the filter mentioned in Section 6.2 can be met by cascading 6 identical FI stages as shown in Figure 6.6. Formed by identical circuit elements, this structure provides a transfer function of

$$\frac{V_{\text{out}}(s)}{V_{\text{in}}(s)} = H(s) = H_1^6(s) \cong \frac{K^6}{\left(1 + s\frac{C}{g_m}\right)^6} \cong \frac{1}{\left(1 + s\frac{C}{g_m}\right)^6}. \tag{6.7}$$

This function gives a DC gain of approximately unity. Assuming that all of the circuit elements ($g_m$ and $C$) in all of the FI stages are identical, the $-3$ dB bandwidth of the LPF in Figure 6.6 will shrink from that of the single FI to

$$f_c = \frac{g_m}{2\pi C}\sqrt{2^{1/6} - 1} = 0.35\frac{g_m}{2\pi C} = 0.0556\frac{g_m}{C}. \tag{6.8}$$

The filter's sensitivity to capacitance and transconductance variations can be found to be

$$S_{g_m,C}^{H(s)} = S_{H_1(s)}^{H(s)} \cdot S_{g_m,C}^{H_1(s)} = 6S_{g_m,C}^{H_1(s)} = 6\left( S_{g_m,C}^{K} - S_{g_m,C}^{\left(1+s\frac{C}{g_m}\right)} \right), \tag{6.9}$$

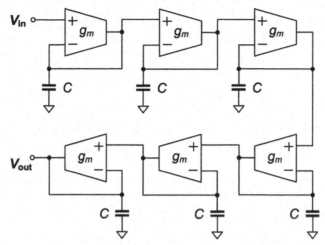

**Figure 6.6**   $6^{th}$-order FI-based LPF.

where $S_{H_{1(s)}}^{H_{(s)}}$ is the sensitivity of the ECG LPF to the transfer function of a single FI, $6S_{g_m, C}^{K}$ is the sensitivity of the ECG LPF's DC gain (numerator of $H_{1(s)}$) to capacitance and transconductance variations and $6S_{g_m, C}^{\left(1+s\frac{C}{g_m}\right)}$ represents the sensitivity of the denominator of $H_{1(s)}$ to capacitance and transconductance variations, respectively.

It can be seen from (6.3) that $K$ equals $g_m R_o (1 + g_m R_o)^{-1}$. This leads to

$$S_C^K = S_C^{\frac{g_m R_o}{1+g_m R_o}} = 0, \tag{6.10}$$

and

$$S_{g_m}^K = \frac{g_m}{K} \cdot \frac{\partial K}{\partial g_m} = \frac{1 + g_m R_o}{R_o} \left( \frac{g_m R_o^2 - (1 + g_m R_o) R_o}{(1 + g_m R_o)^2} \right)$$

$$= \frac{-1}{1 + g_m R_o} \cong 0. \tag{6.11}$$

We can also find that

$$S_{g_m}^{\left(1+s\frac{C}{g_m}\right)} = -S_C^{\left(1+s\frac{C}{g_m}\right)} = \frac{sC}{sC + g_m}. \tag{6.12}$$

Also from (6.8), the sensititvity of $f_c$ to capacitance variation ($S_C^{f_c}$) and to transconductance variation ($S_{g_m}^{f_c}$) can be found as

$$S_{g_m}^{f_c} = \frac{g_m}{f_c} \frac{\partial f_c}{\partial g_m} = \frac{g_m C}{0.0556 g_m} \cdot \frac{0.0556}{C} = 1, \qquad (6.13)$$

and

$$S_C^{f_c} = \frac{C}{f_c} \frac{\partial f_c}{\partial C} = -\frac{C^2}{0.0556 g_m} \cdot \frac{0.0556 g_m}{C^2} = -1, \qquad (6.14)$$

It can be seen that the DC gain of the ECG LPF is insensitive to capacitance and transconductance variations, but the cutoff frequency of the ECG LPF is fully-sensitive to these same variations. Besides, the denominator of $H_{(s)}$ is 6 times more sensitive to capacitance and transconductance variations than the cutoff frequency.

The advantages of this filter topology are the following:

- The linearity of the filter is less susceptible to mismatch than other open-loop structures due to the unity gain feedback in each $1^{st}$-order section.
- The internal node voltage swings are all identical in the passband. Hence, the filter distortionless output swing is maximized [78].
- The contributions of all $g_m$ stages to the overall output noise are almost equal. Hence, the filter output noise is minimized [78].
- Constant group delay can be expected in the passband frequency range.

The drawback of this topology is that the transition band roll-off is less steep compared to other filter types. For a cutoff frequency of 150 Hz to 250 Hz, realizing this filter circuit within a reasonable chip area (i.e., entire filter area not dominated by capacitors only but also by the area of transistors, as both are counted in the cost of fabrication) and obtaining a power consumption of less than 1 nW are feasible.

## 6.4.2 Supply Voltage Requirement, Signal Swing and Tuning

Since all the transconductors are connected in a negative feedback fashion, the input terminals are following each other. In the filter's passband, the transconductor's common mode range (CMR) can be used to approximately define the maximum filter voltage swing. It can be found from Figure 6.5(c) that

$$\text{CMR} \cong V_{\text{ppswing}} = V_{\text{DD}} - V_{\text{ss}} - V_T \left[ 8 + n_p \ln \left( \frac{0.5 I_B}{I_{Sp} \frac{W}{L}_{1,2}} \right) \right], \qquad (6.15)$$

where $I_{Sp}$ is the zero-biased current for a unit transistor ($W = L$) of pMOS devices $M_1$ and $M_2$, and $V_{\text{ppswing}}$ is the desired maximum voltage swing. For a signal swing of less than 50 m$V_p$ and $I_B$ in the sub-nA range, $V_{DD} \cong 0.6$ V should be sufficient for the 0.18-$\mu$m process used in this design. Note that the maximum supply voltage allowed for this process is 1.8 V, 0.6 V equals only one third of that.

Moreover, $I_B$ can be varied to adjust $f_c$ (as $f_c$ is linearly proportional to $g_m$) to compensate for $C$ and $g_m$ variations. From the last term of (6.15) we can see that the voltage required for this tuning (the last term of (6.15) times $V_T$) changes logarithmically with $I_B$. This also implies that $V_{DD} \cong 0.6$ V is sufficient to allow for $f_c$ adjustment to achieve the desired $f_c$ (of 150 Hz or 250 Hz).

### 6.4.3 Signal-to-Noise Ratio

As the filter is formed by cascading 6 identical 1$^{\text{st}}$-order stages with a transfer function defined by (6.7), the integrated output shot noise voltage can be expressed as (assuming the slope factors of nMOS and pMOS to be equal, i.e., $n_p \cong n_n = n$)

$$\overline{v_{\text{noS}}^2} \cong \frac{4nkT}{\pi g_m} \int_0^\infty \left( |H_1(j\omega)|^2 + |H_1^2(j\omega)|^2 + \cdots |H_1^6(j\omega)|^2 \right) d\omega. \quad (6.16)$$

Substituting the transfer function of (6.3) into (6.16), we obtain

$$\overline{v_{\text{noS}}^2} \cong \frac{4nkT}{\pi C}(1.57 + 0.79 + 0.59 + 0.49 + 0.42 + 0.37) \cong \frac{5.4nkT}{C} \quad (6.17)$$

Additionally, the flicker noise can be found by integrating (6.17) from 0.1 Hz to $f_c$ (assuming the flicker noise lies in the filter's passband) and becomes

$$\overline{v_{\text{noF}}^2} = 12 \left( \frac{K_{Fn}}{WL_1} + \frac{K_{Fp}}{WL_3} \right) \ln \left( \frac{5g_m}{\pi C} \right). \quad (6.18)$$

We can approximate the filter's signal to noise ratio (SNR) for input frequencies less than $f_c$ as

$$\text{SNR} = 10 \log \left( \frac{\frac{V_{\text{ppswing}}^2}{8}}{\overline{v_{\text{noS}}^2} + \overline{v_{\text{noF}}^2}} \right). \quad (6.19)$$

It can be seen from (6.17) to (6.19) that large transistor dimensions, capacitor values and a high supply voltage are required to obtain a high SNR, as expected.

### 6.4.4 Supply Noise Rejection and Stability

In the case that $V_{SS}$ is connected to ground, any supply noise that leaks through current source $I_B$ will split and flow into the source terminals of $M_1$ and $M_2$ in Figure 6.5(c) equally (assuming the source impedances of $M_1$ and $M_2$ are identical). Eventually, as they are fully correlated, the split noise sources that flow through transistors $M_1$ and $M_2$ will be minimized at the output node with help of the unity-gain current mirror formed by $M_3$ and $M_4$. For frequencies within the bandwidth of the current mirror, the supply noise can be cancelled out almost completely.

Since the entire ECG LPF is formed by cascading 6 identical stages of FIs without global feedback, the stability of the filter can be inspected from the FI of each stage. Considering Figure 6.5(c) again, by neglecting the gate-drain parasitic capacitances ($c_{gd}$) of $M_3$ and $M_4$, the loop gain of this circuit can be found in the form of a $2^{nd}$-order transfer function with dominant and second poles at $-(CR_o)^{-1}$ and $-g_m(2c_{gs})^{-1}$, respectively, where $R_o$ is the drain-source resistance of $M_2$ and $M_4$ connected in parallel. To maintain circuit stability by guaranteeing $60°$ phase margin, $C \geq 4.4c_{gs}$ is required, assuming $R_o \gg g_m^{-1}$. In practice, as the FI is supposed to be a $1^{st}$-order filter, the design condition of $C \geq 10c_{gs}$ should to be made to avoid the influence of the $2^{nd}$ pole on the required filter transfer function. For this reason, the ECG filter should be always stable.

## 6.5 Design Procedure

In order to design a LPF to meet requirements the the ECG detection mentioned in Section 6.2, the following design methods are recommended.

### 6.5.1 Dynamic Range

From (6.19), we can approximate that SNR $\cong$ DR and $v_{noF} \cong 0$ (we assume $f_{corner}$ is below 1 Hz and $1/f$ noise is negligible), we can calculate the value of capacitor $C$ by approximating that $n_n \cong n_p \cong n \cong 1.4$. To achieve 44 dB DR at room temperature with $V_{ppswing} = 100$ mV, $C = 0.63$ pF is required. This leads to a total capacitance of 3.78 pF for the entire LPF. In the CMOS process we used, 4 metal layers and dual MIM capacitors (each dual MIM capacitor has an extra layer between the top and the $3^{rd}$ metal layers providing higher capacitance per unit area than a regular MIM capacitor) are available.

## 6.5.2 Cutoff Frequency, Bias Current and Power Consumption

After obtaining the value of capacitor $C$, (6.9) can be used to find the value of the transconductance $g_m$ of each transconductor used. Referring to the transconductor in Figure 6.5(c), the transconductance can be found from $g_m = I_B/2nV_T$. For this reason, an $I_B$ of 0.12 nA and 0.21 nA is found at room temperature for cutoff frequencies of 150 Hz and 250 Hz, respectively. As indicated in Figure 6.6, the total current consumption is $6I_B$. In the previous section, we found that a supply voltage of 0.6 V is sufficient for $V_{ppswing} = 100$ mV. Therefore, the power consumption ($P = 6I_B V_{DD}$) of the ECG LPF can be found to be 0.43 nW and 0.76 nW for cutoff frequencies of 150 Hz and 250 Hz, respectively.

## 6.5.3 Tuning

As mentioned in the previous chapter, around 30% cutoff frequency variation can be expected due to process variations [66]. 10% of these variations can originate from the integrated capacitors and 20% stems from transconductance values of the transistors. Although the cutoff frequency is specified to 150 Hz and 250 Hz for this ECG LPF, these values are not strict. As can be seen from the block diagram of the ECG detector in Figure 6.2, the LPF performs anti-aliasing for the consecutive ADC. In general, an anti-aliasing filter with a cutoff frequency sufficiently smaller than the sampling rate of the ADC, 10% variation of $f_c$ is allowed [66]. For this reason, the tuning loop and circuit described in the previous chapter can be applied here as well.

## 6.5.4 Transistor Dimensions

The selection of the transistor dimensions to implement the transconductor in Figure 6.5(c) is influenced by flicker noise, stability and common-mode range. Dimensions of the transistors are set large to produce little flicker noise but they cannot become too large in order not to violate the condition of $C \geq 10c_{gs}$. By means of circuit simulations, we found $W/L_{3,4} = 2.5\,\mu\text{m}/10\,\mu\text{m}$. With this value, $c_{gs3,4}$ of 0.046 pF is obtained. Then, $W/L_{1,2} = 10\,\mu\text{m}/2.5\,\mu\text{m}$ is set to have sufficient headroom for the signal swing as shown in (6.15). For transistors that form $I_B$, their dimensions are larger than the transistor used in the FI circuits to minimize mismatch.

## 6.6 Measurement Results

The filter has been fabricated in the 0.18-$\mu$m CMOS process of AMS. Identical dual MIM capacitors of 0.63 pF are used for each FI stage. The filter's area (bias circuit, transconductors and capacitors) equals 145 $\mu$m 160 $\mu$m. The chip microphotograph is shown in Figure 6.7. The $g_m$ circuit in Figure 6.5(c) was used with transistor dimensions of $W/L_1 = W/L_2 = 10\,\mu$m/2.5 $\mu$m, $W/L_3 = W/L_4 = 2.5\,\mu$m/10 $\mu$m and $W/L_B = 12\,\mu$m/6 $\mu$m. Bias current

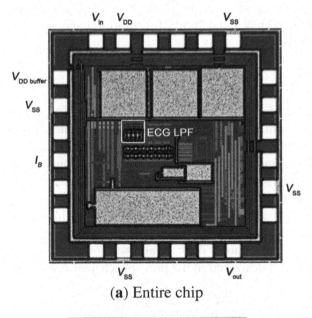

(**a**) Entire chip

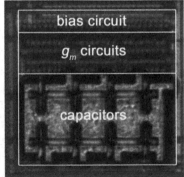

(**b**) Zoomed-in area of the ECG LPF

**Figure 6.7**   Chip photo of the ECG LPF.

$I_B$ was supplied through a simple on-chip current mirror circuit by an external precision current source (Keithley 6430). To drive a large parasitic off-chip capacitor (formed by the pads, package, PCB and measurement probe), a large area PMOS SF was placed at the output of the filter. This SF can affect the filter's linearity so it was biased by a DC current of 10 $\mu$A to be able to drive the load and to have a bandwidth many times higher than $f_c$, thereby minimizing its influence on the filter's linearity. For the nominal test condition, the filter's $V_{DD}$ was set to 0.6 V and the common-mode voltage $V_{CM}$ to 0.32 V. The following results were obtained using a SR785 dynamic signal analyzer.

Figure 6.8 shows the frequency response of the LPF adjusted to cover the ECG frequency range (up to 150 Hz–250 Hz) by $I_B$. For an $I_B$ of 0.14 nA and 0.22 nA (see Figure 6.8(a)), an $f_c$ of 148 Hz and 252 Hz were obtained, respectively. A passband gain of $-0.59$ dB was found for both cases. This value is 0.34 dB below the value obtained from the simulation ($-$ 0.25 dB). In Figure 6.8(b), the phase responses are shown on both log and linear scales. It can be seen that over the filter bandwidth (0–148 Hz and 0–252 Hz), the phase shifts almost linearly and constant group delays (in the passband) of 0.5 ms and 0.85 ms were found for the cases of $I_B = 0.14$ nA and 0.22 nA, respectively. At these values of bias currents, the filter core (excluding the bias circuit) consumes 504 pW and 792 pW power, respectively

Figure 6.9 illustrates the adjustability of the LPF for different values of $I_B$ ranging from 0.1 nA to 10 nA compared with simulations. It can be seen that the filter's cutoff frequency can be adjusted almost linearly over more than two decades (116 Hz–10.5 kHz) and that the largest deviations from the simulated results can be observed for bias currents of 1 nA and 3 nA.

The output noise voltage density for $I_B$ equal to 0.14 nA and 0.22 nA (associated with $f_c$ of 148 Hz and 252 Hz) were measured from 0.5 Hz to 400.5 Hz. The results are plotted along with the simulated results in Figure 6.10. After integrating and referring to the filter's input, we obtain a 187 $\mu$V$_{rms}$ and 175 $\mu$V$_{rms}$ input-referred noise for $I_B = 0.14$ nA and $I_B = 0.22$ nA, respectively. These value are around 35% higher than the value obtained from simulations for both cases.

For the case of $f_c = 252$ Hz ($I_B = 0.22$ nA), the linearity of the filter was tested by applying a 50 Hz sinusoidal input signal and measuring harmonic distortions at the output. We tested linearity of the filter at this frequency to compare with the previously reported ECG filter of [70]. For higher frequencies, the linearity of our filter is expected to be poorer, and better linearity can be expected at lower frequencies, because the loop gain of each FI used is frequency-dependent. For input amplitudes of 50 mV and

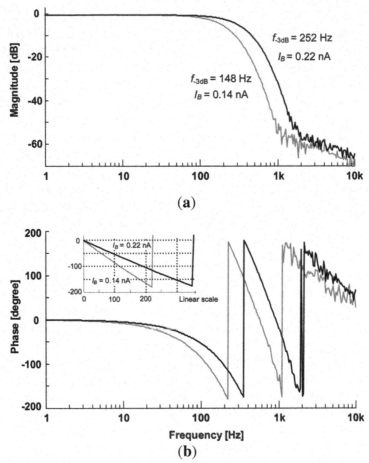

**Figure 6.8**    Measured frequency response. (a) Magnitude response and (b) phase response.

115 mV, respectively, the output power spectra are shown in Figure 6.11. The second harmonic is the main component and it was found at 49 dB and 40.1 dB below the fundamental components for 50 mV and 115 mV input amplitudes, respectively. Note that the 2nd harmonic is dominant here since the signal swing is limited by the CMR of the FI cell, instead of the tanh function that does not produce any even harmonic component. Total harmonic distortion (THD) has been also measured and found at –49 dB and –40 dB for 50 mV and 115 mV input amplitudes, respectively. In the former case ($V_{inp} = 50$ mV), the higher harmonic components are much smaller than the 2nd harmonic component so that the THD is thus mostly influenced by the 2nd harmonic component.

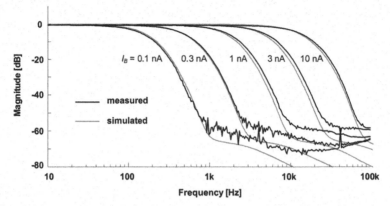

**Figure 6.9** Measured and simulated magnitude response for different bias currents.

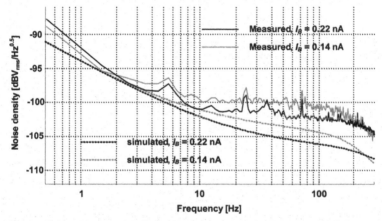

**Figure 6.10** Measured and simulated output noise voltage spectral density.

Relevant characteristics of the filter including DR (SNR at 1% THD) and maximum amplitude for a sinusoidal input signal ($V_{inp}$) were also investigated for different input frequencies ($f_{in}$). The results have been collected and plotted in Figure 6.12. As $f_{in}$ increases, DR (Figure 6.12(c)), and $V_{inp}$ (Figure 6.12(b)) decrease from around 53 dB and 120 mV, respectively, down to slightly higher than 35 dB and 18 mV at the filter's cutoff frequencies for both $f_c = 148$ Hz and 252 Hz. Figure 6.12(a) shows the output amplitude that decreases according to $V_{inp}$ applied as well as the filter's transfer function. This is the result of the reduced loop gain and the widened phase difference between $V_{in}$ and $V_{out}$. At frequencies beyond $f_c$, both DR and $V_{inp}$ rise again since the output amplitude and harmonic components are both attenuated but

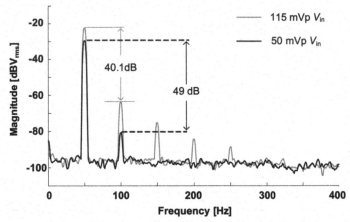

**Figure 6.11**   Measured output voltage spectra of the LPF for different input amplitudes.

at different rates. The higher frequency components are attenuated more than the fundamental component according to the LPF transfer function. Beyond this frequency range, the results are no longer meaningful. Nevertheless, it can be said that the proposed LPF performs worse for $f_{in}$ at around $f_c$ which is a common phenomenon for lowpass filters [79].

In order to test the performance spread of the LPF, ten packaged chips were tested for two cases: magnitude response and harmonic distortion. Figure 6.13 shows the filter's magnitude responses of ten samples for $I_B = 0.22$ nA and $V_{DD} = 0.6$ V (targeting 250 Hz $f_c$). The measured $f_c$ spreads from 233 Hz to 256 Hz, while the range of 237 Hz to 255 Hz was predicted from 10 Monte-Carlo (MC) mismatch simulations. The DC gain was found to be $-0.59$ dB for all ($-0.25$ dB was seen for 10 MC mismatch simulations). It can be seen that the DC gain is less sensitive than the $f_c$ since it is regulated by the large loop gain in the passband. The $f_c$ is sensitive mainly to transistor mismatch in the bias circuit and $g_m$ circuits in the filter core.

Using the same set-up as for the measurement of Figure 6.11, a 50 Hz sinusoidal input was applied to the filter with 115 mV $V_{inp}$ (targeting $-40$ dB THD). The output spectra of ten samples are shown in Figure 6.14. The difference in the output spectrum for each chip can be seen from the 3rd and 5th harmonic components. However, since the main distortion components are the 2nd harmonics which are close to each other for all cases (as they vary less than 0.25 dB), the THD does not vary much from $-40$ dB. The worst case is found at $-39.8$ dB.

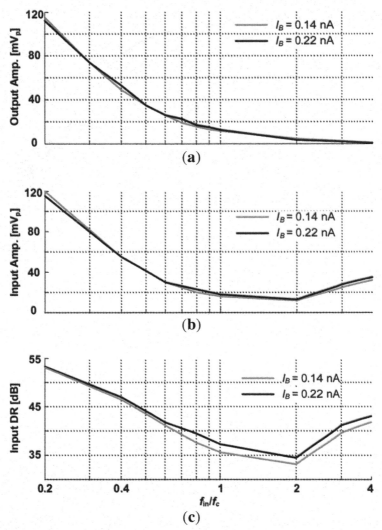

**Figure 6.12** Measured (a) output amplitude, (b) input amplitude and (c) DR at 1% THD versus input frequency.

The main filter characteristics measured for the proposed ECG LPF are summarized for the case of $V_{DD} = 0.6$ V in Table 6.1. Bias currents were set to have cutoff frequences of 150 Hz and 250 Hz, respectively. For measuring the dynamic characteristics (DR, THD, $V_{inp}$ and FoM) of each case, the input frequency was set to $f_{in} = 0.2f_c$. The FoMs were calculated using the formula

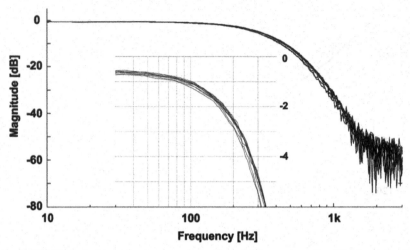

**Figure 6.13**    Measured magnitude responses collected from ten samples. The inset shows the enlarged Y axis (indicated at the right hand side axis of the inset) while its X axis remains the same as for the main plot.

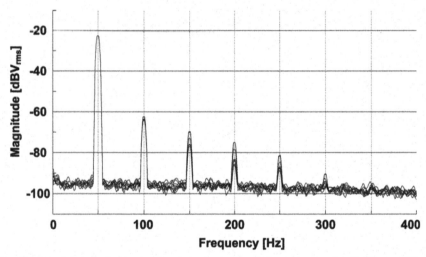

**Figure 6.14**    Measured output voltage spectra collected from ten samples for $V_{inp}$ = 115 mV, $f_{in}$ = 50 Hz, $V_{DD}$ = 0.6 V and $I_B$ = 0.22 nA.

adapted from [80] as shown in 6.1). It can be seen from the table that the proposed LPF provides FoMs for both cases ($f_c$ = 148 Hz and $f_c$ = 252 Hz) approximately three orders of magnitude better than that of the state-of-the-art ECG filter in [70].

**Table 6.1** ECG LPF performance summary for $V_{DD} = 0.6$ V

| $I_B$ | 0.14 nA | 0.22 nA |
|---|---|---|
| $P$ [nW] | 0.504 | 0.792 |
| $f_c$ [Hz] | 148 ; [‡]150 | 252 ; [‡]250 |
| THD@50 mV$_p$ [dB] | –51.3 (30 Hz $f_{in}$) | –49 dB (50 Hz $f_{in}$) |
| [†]$V_{inp}$ [mV] | 120 (30 Hz $f_{in}$) | 115 (50 Hz $f_{in}$) |
| DC gain [dB] | –0.59 ; [‡]0 | –0.59 ; [‡]0 |
| *Output noise voltage | 166 $\mu V_{rms}$ | 155 $\mu V_{rms}$ |
| Input referred noise | 178 $\mu V_{rms}$ | 166 $\mu V_{rms}$ |
| [†]DR [dB] | 53.6 ; [‡]44 (30 Hz $f_{in}$) | 53.8 ; [‡]44 (50 Hz $f_{in}$) |
| **FoM [J] | $10.5 \times 10^{-15}$ | $9.7 \times 10^{-15}$ |

*Integrated from 0.5 Hz to 400.5 Hz, [†]At 1% THD, **calculated using (6.1), [‡]targeted value.

**Table 6.2** ECG filter performance at different $V_{DD}$ and $I_B = 0.22$ nA

| $V_{DD}$ [V] | 0.5 | 0.6 | 0.7 |
|---|---|---|---|
| $V_{CM}$ [V] | 0.26 | 0.32 | 0.37 |
| $P$ [nW] | 0.66 | 0.79 | 0.92 |
| $f_c$ [Hz] | 246 | 252 | 250 |
| DC gain [dB] | –0.63 | –0.59 | –0.55 |
| *$V_{inp}$ [mV] | 75 | 115 | 150 |
| *DR [dB] | 49.6 | 53.8 | 55.7 |
| **FoM [J] | $9.2 \times 10^{-15}$ | $9.7 \times 10^{-15}$ | $11.1 \times 10^{-15}$ |

*Tested at $f_{in} = 50$ Hz and 1% THD, **calculated using (6.1).

**Table 6.3** Performance comparison with other designs

| Ref., Year | [67], 2000 | [68], 2002 | [69], 2005 | [70], 2009 | This Work, 2012 |
|---|---|---|---|---|---|
| $V_{DD}$ [V] | ±1.5 | ±1.35 | ±1.5 | 1 | 0.6 |
| CMOS tech. | 0.8-$\mu$m | 1.2-$\mu$m | 0.35-$\mu$m | 0.18-$\mu$m | 0.18-$\mu$m |
| $N$ | 6 | 2 | 2 | 5 | 6 |
| $f_c$ [Hz] | 2.4 | 0.3 | 37 | 250 | 252 |
| THD [dB] | –50 | –45 | –48.5 | –49.7 | –40 |
| DR [dB] | 60 | 70.5 | 57 | *50 (40.3) | 53.8 |
| $P$ [W] | $10 \mu$ | $8.18 \mu$ | $11 \mu$ | 453 n | 792 p |
| Area [mm$^2$] | 1 | 0.06 | 0.25 | 0.135 | 0.023 |
| **FoM [J]: (6.1) | $1.2 \times 10^{-8}$ | $1.9 \times 10^{-7}$ | $1 \times 10^{-9}$ | *7.3 (8.99) $\times 10^{-12}$ | $9.3 \times 10^{-15}$ |

*It is indicated in Table 6.3 of [3] that DR = 50 dB. This number is incorrect (see Section V of [3]), the correct value is shown here between brackets. Hence, the correct number of FoM is also shown between brackets.

For different values of $V_{DD}$, the filter performance has been also examined. The results are shown in Table 6.2. Common-mode voltages were set to give the highest $V_{inp}$. As expected, the signal-swing related performance including $V_{inp}$ and DR increase with $V_{DD}$. But as the $P$ varies proportionally with $V_{DD}$, according to (6.1), the FoM of all cases are comparable.

Table 6.3 provides relevant characteristics of LPF designs for biomedical applications collected from journal articles (from 2000 to 2012) with measurement results. The filters of [67–70] provide greater DR and better THD than this design. However, the proposed filter can operate from $V_{DD} <$ 1 V, consumes the least power and occupies the smallest area. As a result, the FoM of the proposed ECG LPF is superior to the other existing LPFs by almost three orders of magnitude.

## 6.7 Conclusions

The concept of single-branch filter in Chapter 5 has been extended to the design of an ECG LPF to obtain a major FoM improvement. This achievement comes from using a compact circuit structure with local negative feedback. The measurement results reported confirm the LPF performance in low-voltage and very low-power environments. As the cutoff frequency can be adjusted over two decades, the LPF cannot only be used for ECG application, but also for other types of neural recording systems that are based on a similar concept as the one depicted in the block diagram of Figure 6.2.

# Part III

# Very Low-Frequency Filtering and Large-Swing Multiplication

# 7

---

# Transconductance Reduction Technique for VLF Filters

---

## 7.1 Introduction

Driven by the requirements for very small size and very low-power consumption of portable and implantable medical devices, sub-threshold $Gm - C$ filters have been widely used for filtering low frequency biomedical signals [81]. Conducting a very low-current in subthreshold operation results in a very low transconductance for a single transistor that helps in designing a very low-frequency (VLF) $G_m - C$ filter.

Unfortunately, generating very low bias currents on chip cannot be achieved reliably because of the poor matching between transistors operating in weak inversion [82]. It can be seen from Chapters 5 and 6 that the reliable center and cutoff frequencies of the bandpass and lowpass filters, respectively, are minimally located around 100 Hz from bias currents of $\approx 0.2$ nA. In order to obtain a lower center/cutoff frequency without reducing the bias current further, a circuit technique that reduces the transconductance of any $G_m$ cell employed is required [83].

This chapter investigates transconductance reduction techniques and proposes a more reliable realization that provides a modular design and a cutoff frequency ranging from 100 Hz down to 1 Hz in a reasonable chip area. The proposed transconductor circuit can be formed by compact identical circuit cells and offers both linear range expansion and transconductance reduction, which makes it suitable for very low-frequency, large-dynamic range, low-power, tunable $G_m - C$ filters.

Section 7.2 gives a review of existing techniques for transconductance reduction. The proposed technique with relevant performance analyses will be presented in Section 7.3. An application to the design of a lowpass filter (LPF) with tunable cutoff frequency is shown in Section 7.4. Section 7.5 provides measurement results and a performance comparison between the

107

proposed circuit and previously reported designs. Conclusions are given in the last section.

## 7.2 Review of Transconductance Reduction Techniques

In this work, we classify the transconductance reduction technique into three categories: 1) current attenuation: Figure 7.1(a), 2) current cancellation: Figure 7.1(b) and 3) voltage attenuation: Figure 7.1(c). In most of the cases, each $G_m$ cell is built from an ordinary differential pair circuit such as the one shown in Figure 7.2. For subthreshold operation (weak inversion saturation), the transconductor of Figure 7.2 provides a nonlinear transfer according to

$$I_{\text{out}} = I_B \tanh \left( \frac{V_{i+} - V_{i-}}{2n_p V_T} \right), \tag{7.1}$$

where $n_p$ and $V_T$ are the subthreshold slope factor of pMOS devices $M_1$ and $M_2$, and the thermal voltage, respectively.

The circuit of Figure 7.2 is considered the main source of distortion for the following reasons:

- In the case of Figure 7.1(a), input voltage $V_{\text{in}}$ is applied to the $G_m$ cell and will be subsequently converted into a current before being attenuated by current attenuator $K$, which can be formed by current mirror circuits [84]. Harmonic components are mainly generated by the $G_m$ cell, and since $K$ provides linear current scaling, the ratios of fundamental and harmonic components are maintained. Hence, in this case, the overall transconductance is reduced successfully to $K \cdot G_m$, but the harmonic distortion is unfortunately preserved. Moreover, additional noise contributed by $K$ is unavoidable, which degrades the circuit dynamic range even more.

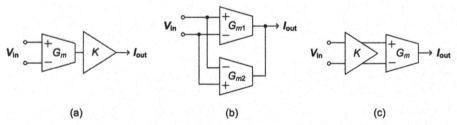

(a)                    (b)                    (c)

**Figure 7.1**   Transconductance reduction techniques. (a) Current attenuation. (b) Current cancellation. (c) Voltage attenuation.

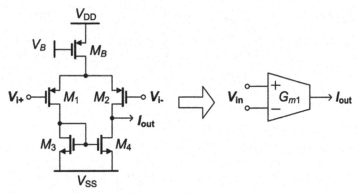

**Figure 7.2** Basic subthreshold transconductor.

- Current cancellation, as shown in Figure 7.1(b) provides transconductance reduction by creating an extra current generated by $G_{m2}$ that is subtracted from the main output current from $G_{m1}$. Effectively, the overall transconductance is reduced to $G_{m1} - G_{m2}$, which (theoretically) can be made extremely small. In practice, the subtraction mechanism is limited by transistor mismatch, and again, more noise is added [85].
- To extend the linear input range of the transconductor as well as to reduce the transconductance, the voltage attenuation technique shown in Figure 7.1(c) is employed. Voltage attenuator $K$ can be made linear by a capacitive dividing network [86] [59]. The overall transconductance in this case equals $K \cdot G_m$, as was also the case when employing current attenuation, but now the linear range is wider without additional noise contribution (only if a noiseless attenuation is used). However, DC offset becomes a problem and the operating point must be set accurately by additional circuitry for this technique to become effective [59]. Moreover, the capacitive dividing network requires a large chip area.

Elaborating further on the voltage attenuation method, we present a transconductance reduction technique in the next section which relaxes the design of bias circuits and provides modularity to the overall design.

## 7.3 Wide-Linear Range Low-$G_m$ Transconductor

This section presents the concept of the transconductance reduction technique and relevant characteristics of the transconductor are analyzed.

### 7.3.1 Concept

Figure 7.3 shows a macro-model of the proposed transconductor. It is composed of an $N$-stage cascade of linear voltage attenuators (attenuation by a factor of 3 in each of $N$ stages) and an ordinary, nonlinear subthreshold transconductor, $G_m$ (Figure 7.2).

By doing so, input voltage $V_{in}$ will be attenuated by a factor of $3^N$ at the input port of $G_m$. Since the amplitude of the voltage appearing at the input port of the nonlinear element becomes smaller, this mechanism effectively leads to linear input range extension as well as transconductance reduction by a factor of $3^N$. The linear attenuation is realized by the network shown in Figure 7.4.

The attenuation network contains four identical nonlinear transconductors, $G_{m1}$ to $G_{m4}$. All of them use local negative feedback. $G_{m1}$ and $G_{m4}$ act as the input transconductors that supply nonlinear current $I_x$ passing through a nonlinear network formed by the back-to-back connection of $G_{m2}$ and $G_{m3}$. With this type of feedback connection, the current flowing through the network of $G_{m2}$ and $G_{m3}$ ($I_x$) is dependent on the voltage across it ($V_3-V_4$). It can also be seen that $G_{m2}$ and $G_{m3}$ together are acting as a nonlinear resistive load for $G_{m1}$ and $G_{m4}$. Assuming that the voltage-to-current relationship of each transconductor is a monotonically increasing nonlinear function with odd-symmetry, i.e.,

$$I_{out} = f\left(V_+ - V_-\right) = -f\left(V_- - V_+\right),\qquad(7.2)$$

we have

$$I_x = f\left(V_1 - V_3\right) = f\left(V_3 - V_4\right) = f\left(V_4 - V_2\right).\qquad(7.3)$$

From (7.3), we can derive:

$$V_1 = 2V_3 - V_4,\qquad(7.4)$$

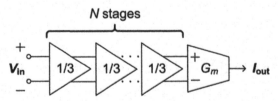

**Figure 7.3**  Proposed transconductor design concept.

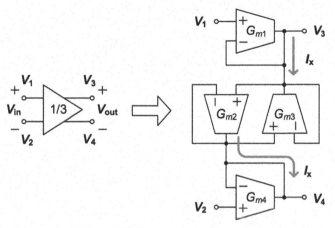

**Figure 7.4**   Linear voltage attenuator.

and

$$V_2 = 2V_4 - V_3. \tag{7.5}$$

Subtracting (7.5) from (7.4), the following linear relationship between input and output voltages is found

$$V_{\text{out}} = V_3 - V_4 = \frac{V_1 - V_2}{3} = \frac{V_{\text{in}}}{3}. \tag{7.6}$$

It can be seen that this linear attenuation can be built from a network of any nonlinear transconductors that comply with (7.2). In practice, both hyperbolic sine and hyperbolic tangent subthreshold tranconductors may be used to implement this concept.

### 7.3.2 Noise and Dynamic Range

Although the input linear range is enlarged, this unfortunately also holds for the input-referred noise. Consider the attenuator circuit shown in Figure 7.5 where each transconductor is assumed to have an output noise current with a power spectral density of $S_{in}(f)$, the power spectral density of the output noise voltage can be found to be

$$
\begin{aligned}
S_{vno}(f) &= S_{vnoP}(f) + S_{vnoN}(f) \\
&= \frac{S_{in1}(f) + S_{in2}(f) + S_{in3}(f) + S_{in4}(f)}{3^2 G_m^2} = \frac{4}{9} \frac{S_{in}(f)}{G_m^2},
\end{aligned} \tag{7.7}
$$

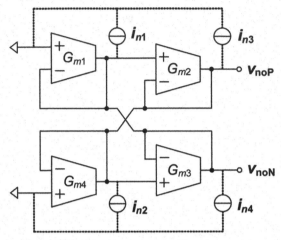

**Figure 7.5**    Linear attenuator with noise sources.

where $S_{vnoP}$ and $S_{vnoN}$ are the output voltage noise spectral densities at nodes $V_3$ and $V_4$, respectively.

For $N$ cascaded stages, we have

$$S_{vtotal}(f) = \sum_{j=0}^{N-1} \frac{S_{vno}(f)}{3^{2j}}. \qquad (7.8)$$

In contrast to the case of a cascade of voltage amplifiers, the majority of noise in this case is generated from stage $N$ instead of the first stage. The noise in (7.8) appears at the input of the main transconductor and will finally be converted into an output noise current in addition to the noise generated from the main transconductor itself. The output noise current spectral density then becomes

$$S_{inout}(f) = S_{in}(f) + \frac{4}{9}\left(S_{in}(f) + \frac{S_{in}(f)}{9} + \frac{S_{in}(f)}{81} + \cdots\right) \qquad (7.9)$$
$$\cong 1.5 S_{in}(f).$$

This leads to a slightly more than 33% reduction in dynamic range compared to the basic transconductor,

$$DR_{proposed} \cong \frac{DR_{basic}}{1.5}. \qquad (7.10)$$

### 7.3.3 Current Consumption and Circuit Complexity

The transconductance reduction in this case is obtained not only by sacrificing the DR but also by increasing the power consumption and the circuit complexity. Since each transconductor operates in class A and consumes a bias current $I_B$, the total current consumption is $(4N + 1) I_B$. For complexity, we count the circuit elements $G_m$ and obtain a total number of elements of $4N + 1$. Both the current and complexity increase by the same factor of $4N + 1$ to obtain transconductance reduction $K$ of $3^N$.

Now it is clear that to save chip area occupied by on-chip capacitors using this technique we need to sacrifice DR and power. Moreover, if $N$ is large enough, the area occupied by the transconductor can be dominant. For this reason, the minimum $N$ can be determined by the minimum current that can be reproduced reliably on chip in practice.

## 7.4 Filter Design and its Measured Results

### 7.4.1 Butterworth 2$^{nd}$-order LPF

In this work, a pseudo differential 2$^{nd}$-order LPF with a transfer function of

$$\frac{V_{\text{out}}(s)}{V_{\text{in}}(s)} = \frac{G_{m1}G_{m2}}{4C_1C_2} \cdot \frac{1}{s^2 + s\frac{G_{m2}}{2C_2} + \frac{G_{m1}G_{m2}}{4C_1C_2}}, \qquad (7.11)$$

its basic topology shown in Figure 7.6, is employed [87, 88]. The differential structure is chosen to widen the signal swing as well as the filter's dynamic range. As the output terminal is fed back directly to the inverting terminals of $G_{m1}$ and $G_{m2}$, a precise following behavior between $V_{\text{in}}$ and $V_{\text{out}}$ is obtained in the filter's passband due to the filter's large loop-gain. This mechanism attains good linearity for input frequencies well below the filter's cutoff frequency [87].

By setting $G_{m1} = G_{m2} = G_m$, we will have a unity passband gain $K \cong 1$, cutoff frequency $f_c = G_m/4\pi\sqrt{C_1C_2}$ and quality factor $Q = \sqrt{C_1/C_2}$.

### 7.4.2 Supply Voltage Requirement and Signal Swing

The differential input voltage of each transconductor in the filter topology in Figure 7.6 is forced by the filter's loop-gain to be very small at frequencies much lower than $f_c$. The common-mode range (CMR) of the transconductor in Figure 7.2 can be used to approximate the maximum filter voltage swing ($V_{\text{ppswing}}$) instead of the transconductor's linear range. Referring to the

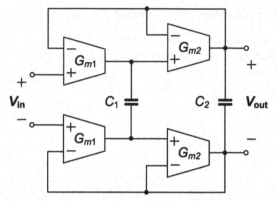

**Figure 7.6** Pseudo differential $2^{\text{nd}}$-order lowpass filter.

transconductor in Figure 7.2, the maximum ($V_{\text{max}}$) and minimum ($V_{\text{min}}$) input voltages that can be applied are

$$V_{\text{max}} \cong V_{\text{DD}} - V_{\text{SDsatB}} - V_{\text{SG1,2}}$$

$$= V_{\text{DD}} - 4V_T - n_p V_T \ln \left( \frac{0.5 I_B}{I_S \left( \frac{W}{L} \right)_{1,2}} \right), \qquad (7.12)$$

and

$$V_{\text{min}} \cong V_{\text{SS}} + V_{\text{GS3}} + V_{\text{SDsat1,2}} - V_{\text{SG1,2}}$$

$$= V_{\text{SS}} + 4V_T - (n_n - n_p) V_T \ln \left( \frac{0.5 I_B}{I_S \left( \frac{W}{L} \right)_{1,2}} \right), \qquad (7.13)$$

assuming that all transistors are operated in the weak inversion saturation regime.

Therefore, CMR and $V_{\text{ppswing}}$ can be found from

$$\text{CMR} \cong V_{\text{ppswing}} = V_{\text{max}} - V_{\text{min}}$$

$$= V_{\text{DD}} - V_{\text{SDsatB}} - V_{\text{SS}} - V_{\text{GS3}}$$

$$\cong V_{\text{DD}} - V_{\text{SS}} - V_T \left[ 8 + n_p \ln \left( \frac{0.5 I_B}{I_S \left( \frac{W}{L} \right)_{1,2}} \right) \right], \qquad (7.14)$$

where $I_s$ is the zero-biased current for a unit transistor (obtained from conditions $V_{GS} = 0, \mid V_{DS} \mid \geq 4V_T$ and $W = L$) of pMOS devices $M_1$ and $M_2$, $n_n$ is the slope factor of nMOS, and $V_{SDsat}$ is the minimum voltage required to keep a transistor in weak inversion saturation (typically, $V_{SDsat} \cong 4V_T$). After selecting $I_B$ for the required values of $f_c$, supply voltage levels $V_{DD}$ and $V_{SS}$ can then be chosen to satisfy the required filter's voltage signal swing at frequencies much lower than $f_c$.

It is worth noting that the extended linear range obtained from the circuit in Figure 7.2 and the CMR in (7.14) define the maximum signal swing of the filter. The maximum signal swing of the filter at frequencies around $f_c$ is limited by the one which is the smallest.

### 7.4.3 Design Procedure

We employ the filter topology in Figure 7.6 and the transconductor of Figure 7.3 with attenuators as shown in Figure 7.4 to design a LPF with $f_c$ in the range from 1 Hz to 100 Hz. We keep the chip area low by limiting the total capacitance to 50 pF. To obtain more than 44 dB DR as required for ECG signals (Chapter 6), the following design steps can to help accomplish the requirements.

#### 7.4.3.1 Passband gain

From the $2^{nd}$-order transfer function expressed in (7.15), a Butterworth magnitude response can be obtained by setting $Q$ to $\sqrt{2}$. Hence, condition $C_1 = 2C_2$ is satisfied. As the total capacitance is limited to $C_1 + C_2 \leq 50$ pF, $C_1 = 2C_2 = 32$ pF are chosen. By doing so, we can expect the filter's passband response to be flat from DC to $f_c$ with a gain of 0 dB. Also, a $-40$ dB/decade roll-off should be be obtained. With this setting, the proposed LPF in Figure 7.6 can also be seen as a differential voltage follower in which the output voltage follows the input voltage closely in the passpand. This following mechanism helps to improve the linearity of the LPF, since the voltage across the input terminals of each transconductor is minimized by the unity-gain feedback connection. As a consequence, the distortion is minimized as well for frequencies well below $f_c$.

#### 7.4.3.2 Minimum bias current and number of stages

As discussed previously that a bias current of less than 0.1 nA cannot be made precise, the minimum value of bias current $I_B$ is limited to 0.2 nA to alleviate the error from transistor mismatch in current mirrors. For this reason, we need to find the value of $G_m$ required first, and then the number of stages ($N$). As $G_m = 2\pi f_c \sqrt{C_1 C_2}$ , values in the range of 0.14 nA/V to 14 nA/V result for the values of $f_c$ required. For $I_B = 0.2$ nA, $n_p \cong 1.6$, $V_T = 26$ mV and transconductance $g_m = 1/(2n_p V_T)$ for the transconductor used (Figure 7.2), the ratio $g_m/G_m \cong 8 = 3^N$ is obtained. Thus, $N = 2$ is chosen. The total bias current required for the entire LPF equals $28I_B$.

### 7.4.3.3 Recovering DR

As the modular transconductors used in this design are similar to the one used in Chapter 6, the same transistor dimensions ($W/L_{1,2} = 10\,\mu\text{m}/2.5\,\mu\text{m}$ and $W/L_{1,2} = 2.5\,\mu\text{m}/10\,\mu\text{m}$) are used here as well. In this frequency range, the flicker noise can be comparable to the shot noise in weak inversion. This degrades the filter DR further in addition to the 30% DR reduction from the attenuator circuit. The loss in DR can be recovered by the following actions:

- Using a differential configuration. By doing so, approximately 6 dB DR improvement should be obtained. Another benefit is a 50% reduction in the size of capacitors used, for the same cutoff frequency. However, the power consumption needs to be double.
- Increase $V_{\text{DD}}$ from 0.6 V (as used for the ECG LPF in Chapter 6) to 1 V to allow for a higher common-mode range of the transconductor used, thereby allowing larger signal voltage swing (see (7.14)). This also leads to more power consumption.

### 7.4.3.4 Bias circuit

As we can see that in total 28 bias current branches are required for the LPF, building a current mirror circuit to supply the bias current to each transconductor is recommended. A large transistor area is selected to minimize mismatch between the transistors used. Also an interdigitated layout is applied. For these reasons, we set $W = 24\ \mu\text{m}$ and $L = 6\ \mu\text{m}$ for the transistors that perform current mirroring in the bias circuit. In this case, approximately 35% of the chip area is occupied by the bias circuit (see Figure 7.7(b)).

### 7.4.4 Measurement Results

The proposed filter has been fabricated in AMS $0.18 - \mu\text{m}$ CMOS technology. In this process, dual metal-insulation-metal (MIM) capacitors (the MIM capacitor with the extra layer to enhance the capacitance density) are available and they were utilized in this design. The transconductors shown in Figure 7.6 are implemented by the modular approach described in Section 6.3 for $N = 2$

(attenuating factor of 9). The total current consumption of the proposed filter core is $36I_B$. The output voltages are buffered via source followers. Figure 7.7 shows the photograph of the chip, which contains the transconductor circuits, bias circuits and dual MIM capacitors $C_1$ and $C_2$. The capacitors overlap some of the active area on the IC. This saves some space and, as a result, the total chip

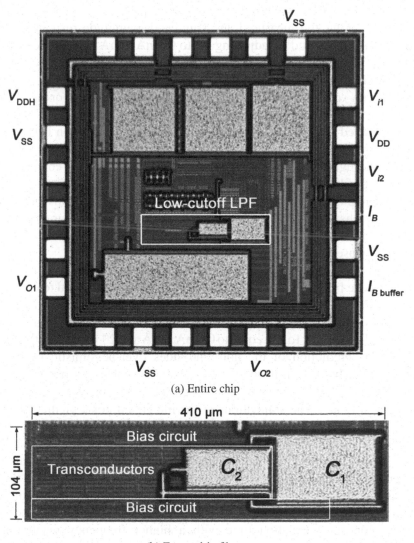

(a) Entire chip

(b) Zoomed-in filter area

**Figure 7.7**   Chip photo.

area is only 0.043 mm$^2$. A Keithley 6430 precision current source has been used to supply DC bias current to the on-chip current mirror that distributes bias currents to the transconductors in the filter core. The following results were measured using an SR785 dynamic signal analyzer under the condition of $V_{SS} = 0$ V, $V_{DD} = 1$ V and $V_{SS} = 0$ V, $V_{DD} = 1.8$ V for the filter core and the source follower buffers, respectively. The differential input voltage has a common-mode level ($V_{CM}$) of 0.5 V.

Figure 7.8 shows the measured and simulated magnitude responses of the proposed LPF obtained by adjusting $I_B$ to 0.2 nA, 2 nA and 20 nA. This results in a measured $f_c$ of 1 Hz, 11.2 Hz and 107 Hz while the simulated $f_c$ is 1.5 Hz, 12 Hz and 110 Hz at the 3 respective bias points set by $I_B$. A measured passband gain ($K$) of $-0.89$ dB can be observed for $I_B = 0.2$ nA. For the case of $I_B = 2$ nA and $I_B = 20$ nA, the same value of $K = -0.62$ dB is obtained. It can be observed for the case of $I_B = 0.2$ nA that the deviation between the simulation and measurement resutls is more than the error for the results obtained for the higher bias currents. This might be due to the diode-connected transistors in the filter circuit (i.e., $M_3$ of Figure 7.2 and some transistor in the bias circuit) are pushed out of their weak inversion saturation at this current level for this realistic case, but they remain in weak inversion saturation for the case of simulation. As a consequence, the drain-source resistances of the transistors decrease as well as the filter's loop gain.

The measured and simulated output noise power spectral densities of the filter for the same biasing conditions and cutoff frequencies as used to obtain the results of Figure 7.8 are shown in Figure 7.9. From these results, the input-referred noise (IRN) voltages, integrated from 1 mHz to $4f_c$ are 133 $\mu V_{rms}$, 234 $\mu V_{rms}$ and 276 $\mu V_{rms}$ for $I_B$ equal to 0.2 nA, 2 nA and 20 nA, respectively. The simulation prediction of the noise provides over-pessimistic results. The flicker noise obtained from simulations covers the entire passband for all cases while the noise corner frequencies observed from the measurements are below $f_c$ for all cases. The simulations of the integrated noise predict more than 60% higher noise than the results obtained by the measurements (see Table 7.1). This might stem from that the flicker noise coefficient in the transistor model provided by the foundry is over-pessimistic. It can also be noticed from Figure 7.9 that for frequencies lower than 2 Hz, the noise densities for all cases converge. This is most probably a measure-ment artifact. To measure the noise behavior in this frequency range more precisely, a longer observation time with a higher frequency resolution is required.

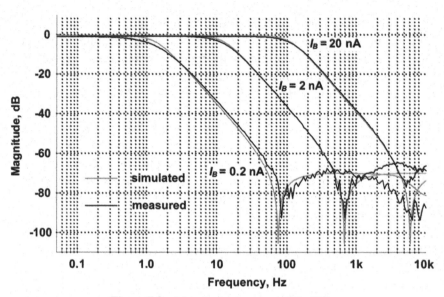

**Figure 7.8** Magnitude responses of the LPF.

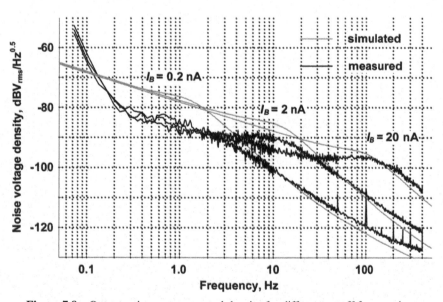

**Figure 7.9** Output noise power spectral density for different cutoff frequencies.

The linearity of the filter has been tested by applying a sinusoidal input voltage with input frequency $f_{in} = f_c$ and observing the filter's output spectrum. For different input amplitudes, the total harmonic distortion (THD) has been collected and plotted as shown in Figure 7.10. For $I_B$ of 0.2 nA, 2 nA and 20 nA (corresponding to $f_c$ of 1 Hz, 11.2 Hz and 107 Hz, respectively), the measured input amplitudes that induce 1% THD are 520 mV, 650 mV and 796 mV, respectively, while 520 mV, 700 mV and 760 mV, respectively, are obtained from simulations. Taking into account the input referred noise voltages mentioned in the previous paragraph, input dynamic ranges of 68.8 dB, 66.2 dB and 66.2 dB are achieved, respectively. With the noise obtained from the simulations, it follows that the simulated DR is maximally 7 dB different from the measured DR.

Table 7.1 presents a performance summary of the filter tested at different cutoff frequencies. The DR of the filter is larger than 66 dB for the entire range of $f_c$ tested. The power consumption of the filter varies with $f_c$, and the filter consumes only 720 nW at the highest $f_c$. For the figure of merit (FoM) defined by

$$\text{FoM} = \frac{P}{M \cdot f_c \cdot \text{DR}}, \tag{7.15}$$

where $P$ is the power consumption and $M$ is the filter order, we achieve a FoM between 1.9 and 3.2 fJ for all cases.

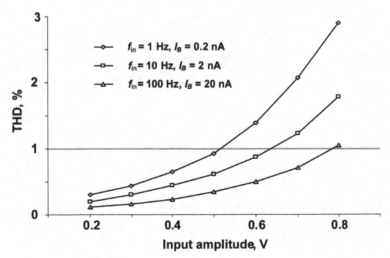

**Figure 7.10**    Total harmonic distortion versus input amplitude.

**Table 7.1** Peformance summary

| Filter Characteristic | Measured | | | Simulated | | |
|---|---|---|---|---|---|---|
| $I_B$ [nA] | 0.2 | 2 | 20 | 0.2 | 2 | 20 |
| $K$ [dB] | −0.89 | −0.62 | −0.62 | −0.35 | −0.31 | −0.29 |
| $f_{-3dB}$ [Hz] | 1 | 11.2 | 107 | 1.5 | 12 | 110 |
| *Output noise [$\mu V_{rms}$] | 120 | 209 | 257 | 287 | 366 | 400 |
| IRN [$\mu V_{rms}$] | 133 | 224 | 276 | 299 | 379 | 414 |
| $V_p$@1% THD [V] | 0.52@ 1 HZ | 0.65@ 10 Hz | 0.79@ 100 Hz | 0.52@ 1 Hz | 0.7@ 10 Hz | 0.76@ 100 Hz |
| DR [dB] | 68.8 | 66.2 | 66.2 | 61.8 | 62.2 | 62.3 |
| †Power [nW] | 7.2 | 72 | 720 | 7.2 | 72 | 720 |
| FoM [fJ] | 0.48 | 0.77 | 0.8 | 1.6 | 1.8 | 1.9 |

*Integrated from 10 mHz to $4f_{-3dB}$.

†excluding bias circuit.

Table 7.2 compares measurement results for the proposed LPF with previously reported very low-frequency LPFs. The proposed filter outperforms the others in terms of chip area and FoM. Regarding power consumption, this filter does not consume as little as the filter proposed in [89]. However, the filter of [89] is not performing entirely in the continuous-time domain. Instead, a switching technique is utilized to synthesize a very large time constant. Compared with the 2nd-order Butterworth structure in [88] (using the approach most similar to our design), our filter provides a tuning range that is one decade larger while the power consumption of our filter is more than three orders of magnitude lower.

**Table 7.2** Measured performance summary and comparison

| Reference | [67] | [90] | [91] | [92] | [88] | [89] | This Work |
|---|---|---|---|---|---|---|---|
| Year | 2000 | 2003 | 2004 | 2006 | 2007 | 2011 | 2012 |
| CMOS tech. [$\mu$m] | 0.8 | 0.5 | 0.8 | 0.8 | 0.35 | 0.35 | 0.18 |
| Order ; topology | 6 ; SE | 1 ; SE | 1 ; SE | 1 ; SE | 2 ; SE | 1 ; SE | 2 ; DF |
| Chip area [mm²] | 1 | 0.035 | 0.1 | 0.2 | 0.336 | 0.07 | 0.043 |
| Capacitance per pole [pF] | 5 | 15 | 70 | 50 | 52.5 | 40 | 24.18 |
| Minimum $f_c$ [Hz] | 2.4 | 0.18 | 0.1 | 0.3 | 1.5 | 0.02 | 1 |
| $P$ [W] | 10 $\mu$ | 77.4 n | 230 n | 113 n | 165 $\mu$ | 5 n | 7.2 n |
| $V_{DD}$ [V] | 3 | 3 | 3 | – | 3.3 | 1 | 1 |
| THD [%] | 1 | – | – | – | 1 | 0.96 | 1 |
| DR [dB] | 60 | 61.4 | – | 65.4 | 60 | 63.8 | 68.8 |
| FoM [fJ] | 694 | 312 | – | 109 | 5500 | 104 | 0.48 |

SE = single ended, DF = differential.

## 7.5 Conclusions

A modular transconductance reduction technique has been developed to successfully implement a fully-integrated, very low-frequency $2^{nd}$-order Butterworth LPF in AMS 0.18 $\mu$m CMOS process. By using an ordinary sub-threshold differential pair transconductor as a basic module, the filter achieves a cutoff frequency tuning range of two decades at a power consumption much lower than one $\mu$W for the highest $f_c$. For lower $f_c$, the power consumption scales down linearly. Based on its FoM and chip area, the proposed filter also provides the best result compared to other existing very low-frequency $G_m - C$ LPFs. For these reasons, the area-efficient filter proposed in this chapter can be useful for very low-frequency biomedical applications.

# 8

## Large-Swing Current Multiplier
## for AP Detection

*"The energy of the mind is the essence of life."*

— Aristotle

## 8.1 Introduction

An action potential (AP) or spike detector is an important part of neural recording implants [93]. The detector performs on-chip data reduction by trying to capture only relevant information (real occurrences of the action potential) from the recorded signals. This data selection is useful for reducing the data rate of the neural sensor wireless transmission, which eventually leads to a reduction in operating power [93]. To locate the real-time occurrences of action potentials in noisy environments, a 'nonlinear energy operator, (NEO)' [94], has been widely used to discriminate between action potentials and the noise in which their energies are considered different [95]. The NEO provides real-time energy detection and takes into account both amplitude and frequency of the signal of interest. The NEO algorithm is described by

$$y(t) = \Psi(x(t)) = \left(\frac{dx(t)}{dt}\right)^2 - x(t)\frac{d^2x(t)}{dt^2}, \qquad (8.1)$$

where $x(t)$ represents the signal of interest in the time domain and $y(t)$ is the real-time energy of $x(t)$.

To examplify that this algorithm involves not only the amplitude, but also the frequency of the signal, one can substitute $x(t) = A\cos(\omega t)$ into (8.1). The result of $\Psi(x(t)) = A^2\omega^2$ will be obtained, where $A$ and $\omega$ stand for the amplitude and angular frequency of $x(t)$, respectively.

123

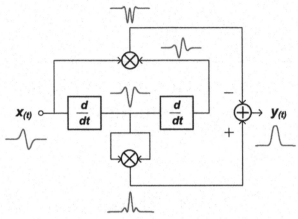

**Figure 8.1**    Nonlinear energy operator (NEO).

It can be seen from (8.1) that the NEO algorithm itself is compact since only differentiation, multiplication, and subtraction are required. Therefore, the use of analog circuits to realize the NEO is being considered in the field of ultra low-power integrated circuit design for implantable devices [96–98].

Equation 8.1 can also be represented in the form of a block diagram as shown in Figure 8.1. It comprises two differentiators, two multipliers, and a subtraction block. The latter operation is simply realized in an analog fashion by representing the signals in term of currents and using a current mirror to perform the subtraction. So the differentiators and the analog multipliers are the main design challenges. The differentiator suppresses low frequency components in $x(t)$ while, on the other hand, high frequency components of $x(t)$, will be enhanced. This is because the variation of the amplitude over time ($\frac{dx(t)}{dt}$ or the slope) of high-frequency signals is faster than that of low-frequency signals. Careful consideration of the frequency content of $x(t)$ and the differentiator circuit time constant, power consumption, and dynamic range, is thus required.

Focusing on the first differentiator at the input, its output signal is supplied to both the lower multiplier and the second differentiator. The output signal amplitude of the second differentiator (identical to the first one) will be amplified again for high-frequency signals that have been amplified by the first differentiator. On the other hand, the amplitude of low-frequency signals that have been suppressed by the first differentiator will be suppressed further by this differentiator. These twice suppressed and amplified signals will be supplied to one of the input signals for the upper multiplier. Now it is clear that the multiplier for the NEO-based spike detector needs to handle a wide signal range (i.e., small amplitudes of low-frequency signals and large amplitudes of high-frequency signals). To keep the operating power low and to maintain a wide input signal range, class-AB operation, current-mode signal processing [99], and MOSFETs operating in weak inversion are considered for the multiplier design.

Based on the well-known exponential characteristics of BJTs or weak inversion MOSFETs, four-quadrant current multiplier circuits have been designed using different principles, e.g., transconductor/conveyor-based [100], and translinear circuit-based [101] current multipliers. Most of them are restricted to class-A operation, which does not allow the input signals swing to exceed the bias currents. To handle large input signal amplitudes, a class-A current multiplier requires large bias currents, which subsequently results in high power consumption. In this chapter, a class-AB four-quadrant analog current multiplier is presented. The proposed multiplier is formed by a dual output current amplifier which is biased by controlled currents generated from a current splitter. Both the amplifier and splitter circuits can be realized from the same basic circuit block, called a sinh transconductor featuring class-AB operation. Therefore, overall class-AB operation for the multiplier is obtained. As a result, the multiplier circuit can be designed to process high input signal amplitudes while its bias current can be kept low. Circuit simulation using a 0.13-$\mu$m transistor model shows that, for a 0.5 nA bias current, input current amplitudes of 5 nA can be applied to the multiplier circuit, and good four-quadrant multiplication is achieved.

## 8.2 Class-AB Current Multiplier

Figure 8.2 shows the block diagram of the proposed multiplier. It comprises three identical current-controlled nonlinear transconductors, $G_o$, $G_A$, $G_B$ and a class-AB current splitter, $K$, that supplies signal-dependent currents $I_A$ and $I_B$ to bias transconductors $G_A$ and $G_B$. Let's assume that the $I - V$ characteristic of each transconductor is a strictly monotonic function described by

$$I_{\text{out}} = I_o f \left( V_+ - V_- \right), \tag{8.2}$$

where $I_o$ is the bias current applied to the bias node of the transconductor. Using (8.2), voltage $V_i$ in Figure 8.2 can be found as

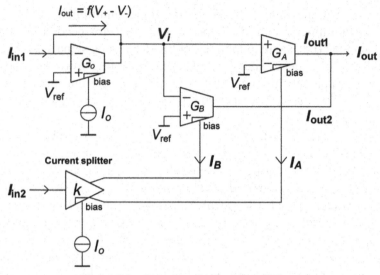

**Figure 8.2**   Current multiplier block diagram.

$$V_i = f^{-1}\left(\frac{I_{\text{in1}}}{I_o}\right) - V_{\text{ref}}. \tag{8.3}$$

Unlike $G_o$ that is biased by DC current $I_o$, $G_A$ and $G_B$ are biased by signal-dependent currents $I_A$ and $I_B$, respectively. Substituting (8.3) into (8.2) for $G_A$ and $G_B$, the output currents of both transconductors are:

$$I_{\text{out1}} = \frac{I_A}{I_o}I_{\text{in1}}, \tag{8.4}$$

and

$$I_{\text{out2}} = -\frac{I_B}{I_o}I_{\text{in1}}. \tag{8.5}$$

It is clear from (8.4) and (8.5) that the monotonic function shown on the right hand side of (8.2), which, in practice, arises from the nonlinear behavior of the semiconductor devices used to implement the transconductor, has disappeared. This results in linear relationships between the input and output currents. Therefore, the output current of the multiplier can be found to be

$$I_{\text{out}} = I_{\text{out1}} + I_{\text{out2}} = \frac{I_A - I_B}{I_o}I_{\text{in1}}. \tag{8.6}$$

In this case, $I_o$ is a constant current but $I_A$ and $I_B$ are generated from input current $I_{in2}$, via the current splitter, according to the following relation

$$I_{in2} = k\,(I_A - I_B)\,.  \tag{8.7}$$

Substituting (8.7) into (8.6), we arrive at

$$I_{out} = \frac{I_{in1} I_{in2}}{k I_o} = A \cdot I_{in1} I_{in2}.  \tag{8.8}$$

It can be seen from (8.8) that a four-quadrant multiplication is obtained from the circuit in Figure 8.2 with a conversion gain of $A = (k I_o)^{-1}$. This allows us to adjust the gain electronically by varying bias current $I_o$.

In comparison with the well-known Gilbert multiplier, the fundamental difference is that the Gilbert multiplier uses a transconductor to convert the input voltage into currents first, and the circuit operation is limited to class A. In this class-AB structure, the input signals are currents and one of them is converted into voltage first, and another one is split into two current components. The similarity between them is that there is a pair of transconductors having their bias currents modulated by one of the input signals, and the output currents of both multiplier circuits are obtained from the summation of the output currents of the transconductor pairs.

In CMOS technology, the multiplier can be designed to operate in class-AB at a power consumption down to a few nW. The exponential behavior of subthreshold CMOS devices is employed to design the transconductors and the current splitter. They are operated from a low supply voltage (less than 1 V) and are described in the next section.

## 8.3 Subthreshold Class-AB Building Blocks

Figure 8.3(a) shows the nonlinear class-AB transconductor which can be directly substituted for $G_o$, $G_A$ and $G_B$ of Figure 8.2. Using the exponential relationship of MOSFETs operating in weak inversion saturation [40], for $V_{SB} = 0$ (source and body terminals are connected), it follows that

$$I_D = I_{D0} \exp\left(\frac{V_{SG}}{n_p V_T}\right),  \tag{8.9}$$

where

$$I_{D0} = I_{S0}\left(\frac{W}{L}\right).  \tag{8.10}$$

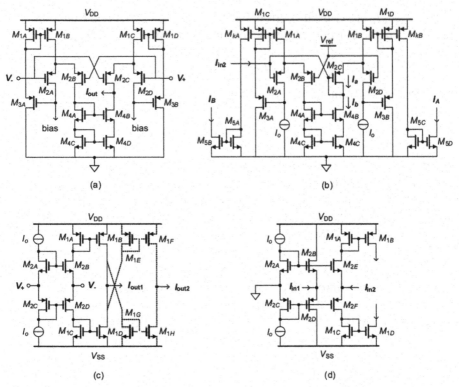

**Figure 8.3**  Circuit building blocks. (a) sinh transconductor. (b) sinh-based class-AB current splitter. (c) Complementary sinh transconductor. (d) Complementary sinh-based class-AB current splitter.

Current $I_{S0}$ is the zero-biased current for a unit transistor ($W = L$) obtained by setting $V_{SG} = 0$, $n_p$ is the subthreshold slope factor of the pMOS and $V_T$ is the thermal voltage. $W$ and $L$ represent the width and length, $V_{SB}$ is the source-bulk voltage, and $V_{SG}$ the gate-source voltage of the transistor.

As the set of cascode transistors $M_{4A}$–$M_{4D}$ forms a unity-gain current mirror, we can use (8.9) to find the input-output relation of each transconductor as [102]

$$I_{\text{out}} = 2I_o \sinh\left(\frac{V_+ - V_-}{n_p V_T}\right) \tag{8.11}$$

This relationship complies with (8.2) and provides class-AB operation. Note that in the case of $V_{SB} \neq 0$ for the set of transistors $M_2$, the bulk effect will not affect the relationship of (8.11) but will modify the value of $I_S$, and for this reason, the transconductor may require a higher supply voltage to maintain the same operating condition.

Figure 8.3(b) shows the current splitter circuit. It is formed by the same circuit topology as the transconductor in Figure 8.3(a). Current mirror circuits $M_{kA} - M_{5A} - M_{5B}$ and $M_{kB} - M_{5C} - M_{5D}$ are inserted to implement the output branches, $I_B$ and $I_A$, respectively. The negative feedback is applied to the input node to allow input current $I_{in2}$ flowing into the translinear loop formed by transistors $M_{2A}$ to $M_{2D}$. After entering the loop, $I_{in2}$ will be split into two components $I_a$ and $I_b$. With help from the unity-gain cascode current mirror $M_{4A}$–$M_{4D}$ and the additional current mirrors $M_{kA} - M_{5A} - M_{5B}$ and $M_{kB} - M_{5C} - M_{5D}$, $I_a$ and $I_b$ can be conveyed to the output by $I_A = k(I_o + I_a)$ and $I_B = k(I_o + I_b)$.

Consider the translinear loop comprising $M_{2A}$ to $M_{2D}$ again, according to the translinear principle [103], the following relationships can be formed.

$$I_{in2} = I_a - I_b = I_A - I_B \tag{8.12}$$

and

$$I_o^2 = I_a I_b. \tag{8.13}$$

The output currents, $I_A$ and $I_B$ will be delivered to $G_A$ and $G_B$ in the full current multiplier circuit, respectively, with scaling factor $k$ defined by the dimensions of $M_1$ and $M_k$.

Other possible ways to design the sinh transconductor and current splitter are shown in Figures 8.3(c) and 8.3(d), respectively. This approach employs complementary devices to implement the same translinear-loop equations as obtained from the circuit in Figures 8.3(a) and 8.3(b). However, they suffer from different subthreshold slopes due to the body effect on the nMOS and pMOS devices, something which is unavoidable in standard CMOS processes [102]. For this reason, we choose the circuits in Figures 8.3(a) and 8.3(b) to validate our design.

## 8.4 Simulations

The multiplier in Figure 8.1 was designed employing the circuit blocks in Figures 8.2(a) and 8.2(b) and simulated using Spectre RF and TSMC 0.13-$\mu$m CMOS model parameters. Transistor widths ($W$) and lengths ($L$) are given in

Table 8.1. $V_{\text{DD}}$ and $V_{\text{ref}}$ were set to 0.65 V and 0.4 V, respectively. A bias current $I_o$ of 0.5 nA was used. The quiescent power of the entire circuit is 12.4 nW.

The transient response illustrating the four-quadrant multiplication of a 5 nA, 2 kHz, sinusoidal current $I_{\text{in1}}$ (this value implies a modulation index (MI), which is the ratio of the signal amplitude over bias current $I_o$, of 10) and a 5 nA, 100 Hz, sinusoidal current $I_{\text{in2}}$ (MI = 10) produced by the proposed multiplier is shown in Figure 8.4. It is seen that the circuit performs the multiplication function well in the time domain for these very high input signal modulation indices.

To examine the circuit linearity, a circuit simulation of the total harmonic distortion (THD) has been performed by varying the amplitude of the 2 kHz sinusoidal $I_{\text{in1}}$ from 1 nA to 5 nA ($2 \leq \text{MI} \leq 10$), keeping $I_{\text{in2}}$ constant at 5 nA. The results shown in Figure 8.5 reveal that at a MI of 10, the proposed multiplier provides approximately $-30$ dB THD.

**Table 8.1**  Transistor dimensions

| MOSFET | $M_{1A-D}$ | $M_{2A-D}$ | $M_{3A-B}$ | $M_{4A-D}$ | $M_{5A-D}$ | $M_{kA-B}$ |
|---|---|---|---|---|---|---|
| $W/L\ [\mu m/\mu m]$ | 4/4 | 1/1 | 1/10 | 2/2 | 2/2 | 2/4 |

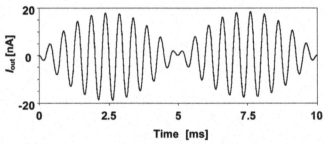

**Figure 8.4**  Simulated transient response of the multiplier. $I_{\text{in1}} = \hat{I}_p \sin(2\pi f_1 t)$ and $I_{\text{in2}} = \hat{I}_p \sin(2\pi f_2 t)$, where $f_1 = 100$ Hz, $f_2 = 2$ kHz and $\hat{I}_p = 5$ nA.

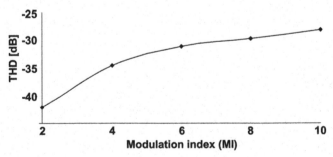

**Figure 8.5**  Simulated total harmonic distortion.

## 8.5 Conclusions

A low-voltage, ultra-low-power CMOS four-quadrant current multiplier has been described. By using a weak inversion sinh transconductor as a basic cell, the first fully class-AB multiplier is obtained. The multiplier processes input signals with a modulation index of 10 while the circuit DC power consumption can be kept very low. It must be noted that under large-signal drive, power consumption will increase. This multiplier is suitable not only for realizing a NEO-based action potential detector, but also for other implantable systems that need to be operated from a very low-supply voltage with very little power consumption.

# 9

# Conclusions and Future Work

This book has described IC design techniques for nanopower analog signal processing in biomedical applications. The work has attempted to serve the requirements of extremely low-power operation and reasonable physical size for wearable and implantable medical devices by maximing the capability of a single MOS transistor. Discrete and continuous-time design techniques have been addressed. The book has also dealt with the nonlinear behavior of conventional circuit building blocks in severe requirements such as the very low cutoff frequency needed to filter very slowly varying bio-potentials, or the very large signal swing demanded by energy-based action potential detection.

## 9.1 General Conclusions

Part **I** of this book presented the investigation of a subthreshold, switched-current memory cell from a feedback point of view. Including switching nonidealities and large-signal characteristics of the transconductor, it has been revealed in Chapter **3** that using a MOS transistor as a $G_m$ cell alone cannot provide sufficient performance, since charge injection and clock-feedthrough effects cannot be suppressed completely. To minimize the performance degradation and enhance circuit dynamic range at very low power consumption, a very large loop gain (as can be obtained from a two-stage circuit), a fully-differential architecture and class-AB circuit operation are required. The design presented in Chapter **4** demonstrated a class-AB SI memory cell that follows on from the findings obtained in Chapter 3. The current-domain memory cell characterized in simulation can handle very large input amplitudes with high dynamic range for a wide range of input frequencies (up to 50 kHz $f_{in}$, 1 MS/s $f_s$). In terms of figure-of-merit, this proposed design has advanced beyond existing designs by more than an order of magnitude.

Unfortunately, an oscillation appears when handling input signal levels that overload the memory cell input due to the perfromance constraint,

$I_{\text{idmax}} \cong 2I_{B2} \sinh\left(\text{acosh}\left(3.125\frac{C_H}{C_1+C_2}\right)\right)$ of the transconductor employed in the two-stage memory cell topology when driven large signal. In addition, the overall circuit complexity is relatively high because the two-stage fully-differential topology requires common-mode feedback circuits. However, its applicability (assuming that the stability problem can be solved by the solution proposed in Section 9.3) in a cochlear implant (CI) speech processor will be discussed in greater detail in Appendix A.

Part **II** of this book established the concept of a single MOS circuit design via the implementation of single-branch filters. Chapter **5** described a single branch BPF that realizes 2 time constants using only 2 MOS transistors and 2 capacitors, and that consumes only one branch of bias current. The concept has been verified by experimental measurement of a $4^{\text{th}}$-order BPF fabricated in a 0.18-$\mu$m CMOS IC technology. The filter outperforms existing BPFs by an order of magnitude in FoM. The power consumption of this filter is a few nW, and the area occupied is smaller than other designs. The center frequency of the filter is also adjustable by more than 6 octaves, covering the audio range required in CI processors. Chapter **6** applied the single-branch concept to implement a LPF that fits the requirements for a portable ECG detector. Experimental results from a $6^{\text{th}}$-order ECG LPF prototype fabricated in a 0.18-$\mu$m CMOS technology show that the design methodology developed in this book reduces filter power consumption by more than two orders of magnitude. The area occupied by the circuit on chip has also been reduced by more than a factor of two compared to existing implementations.

The two filters presented in Part **II** of this book have successfully met the power and size reduction targets required by medical devices. A limitation of these two filters is that they cannot realize complex conjugate poles. Therefore, the transition band roll-off found in their magnitude response is not as steep as for a filter with complex poles.

Part **III** presented techniques for the design of analog circuits to meet the extreme requirements found in biomedical applications. These requirements have been met by applying both negative feedback and large-signal nonlinear cancellation. Chapter **7** discussed a modular technique for realizing a very low cutoff frequency $G_m - C$ LPF. This technique uses the concept of voltage attenuation, which not only reduces the transconductance, but also enlarges the linear input range of the transconductor. The voltage attenuation is linear and realized using a network comprised of a handful of ordinary nonlinear transconductors. Verified experimentally for a $2^{\text{nd}}$-order LPF fabricated in a 0.18-$\mu$m CMOS technology, this design technique achieves very low cutoff frequencies (100 Hz down to 1 Hz) from reasonably high bias currents (20-0.2 nA) and reasonable chip area (0.034 mm$^2$). In comparison with other very low-frequency integrated LPFs reported in the recent literature, this design attains the best FoM and the smallest physical dimensions. Chapter **8** discussed a four-quadrant current multiplier operating in weak inversion that meets the wide input signal swing requirement for an energy-based spike detector. Forming a network of nonlinear circuit blocks including sinh transconductors and a current splitter, the first demonstration of a class-AB four-quadrant analog multiplier is obtained. The correct operation of the multiplier has been verified by circuit simulations using 0.13-$\mu$m CMOS model parameters. Consuming only a few nW of static power, input currents can be as high as 10 times the circuit bias current, while acceptable linearity can be maintained.

## 9.2 List of Achievements

The outcomes of the research work in this book can be summarized as follows:

- A fully-differential, class-AB, current-mode, sample-and-hold circuit operating the CMOS transistor in weak inversion has been designed and verified by circuit simulations. Compared to other designs reported in the recent literature, more than an order of magnitude improvement in FoM is achieved.
- A 4$^{th}$-order BPF suitable for a CI processor has been implemented in the AMS 0.18 $\mu$m CMOS technology. In comparison to other BPFs reported to date, an order of magnitude improvement in the FoM and the smallest occupied chip area are achieved.
- A 6$^{th}$-order LPF dedicated to ECG detection has also been implemented in the AMS 0.18 $\mu$m CMOS technology. Measurement results show that the LPF meets the requirements for ECG recording with approximately 3 orders of magnitude improvement in FoM compared to other recently reported LPFs.
- A fully-integrated, 1–100 Hz, 2$^{nd}$-order $G_m - C$ LPF has been implemented in the AMS 0.18 $\mu$m CMOS technology. The LPF attains the best FoM (by approximately two orders of magnitude improvement) and the smallest chip area in comparison with other very-low frequency filters reported in the literature.
- The initial concept and design of a subthreshold-biased, class-AB four-quadrant multiplier circuit operating in the current-mode has been

introduced in this book. As verified from circuit simulations, the multiplier can handle input current amplitudes as high as 10 times the bias current whilst consuming just a few nW of quiescent power.

## 9.3 Future Work

The following items can be developed further:

- The instability of the SI memory cell at larger input amplitudes reported in Part **I** should be eliminated. This could be accomplished through an adaptive biasing scheme. The input signal modulating the operating point of the main transconductor can be tracked and fed back to set the bias current of the input amplifier. By doing so, the phase margin of the overall circuit's loop gain is expected to be maintained constant across the entire input signal range and a sample-and-hold operation with a DR of more than 80 dB can be expected.
- The current memory cell could be applied to the design of a peak instant detector (see the Appendix A). This can aid the design of an analog CI processor by being able to convey the fine structure of the incoming sound to the stimulator in a very compact and low power fashion.
- For the filter bank section of a CI processor, which is combined with a logarithmic compressor, the BPF reported in Chapter **5** can be utilized. Very low power consumption can be expected.
- The LPF in Chapter **6** should be developed to allow for the realization of complex poles. Instead of only using a FI as a building block of the filter, we should allow a second-order section with complex poles to be a building block as well. As a result of this development, a lower-order LPF with sufficient transition band attenuation should become possible. This can help reducing overall power consumption and chip area further.

# Appendix A

## Phase-Locked Peak-Picking
## Speech Processor

### A.1 Introduction

Recovering human hearing perception via electrical stimulation is the ultimate goal of contemporary bionic ear (BE) or cochlear implant (CI) devices. To some extent this goal has been achieved by the invention of a speech processing strategy called 'continuous interleaved sampling, (CIS)' [104]. It roughly emulates the behavior of the basilar membrane and inner and outer hair cells, and successfully prevents simultaneous interactions between electrodes using fixed-rate interleaved amplitude-modulated stimuli. CIS has been employed as a default strategy in several commercially available CI devices from different manufacturers, i.e., MED-EL GmbH, Cochlear Ltd. and Advanced Bionic Corp. The results obtained from clinical experiments have shown to offer reliable understanding of sentences in quiet environments but poor results obtained for simple melodies. In typical noisy environments, the patients (CI recipients) are still having difficulties to understand both sentences and melodies [105]. These indications imply that the temporal fine structure (TFS: fast varying components of the sound) is not being conveyed to the brain.

In normal mammalian auditory nerve fibers, the spike trains synchronize with the stimulus waveform periodicity up to 5 kHz [106]. Beyond that frequency range, the spike trains are generated randomly [107]. The aforementioned mechanisms are missing in the CIS processor since the pulse stimulation rate is fixed. To gain the perception of tonal languages and music, an effort of realizing a BE processor that imitates the inner hair cells and the auditory nerve behavior more precisely is being considered. For this reason, the 'Hilbert transform, (HT)' has been introduced to the BE processor to extract temporal envelope, instantaneous frequency and phase, and thereby the TFS [108].

Although extraction can be achieved, conveying all of the information to the brain via electrical pulse trains is still a challenge that remains. Besides, performing the HT requires a large computational cost for both digital [109] and analog [110] processors.

This appendix explores some envelope-based processing strategies that require the computationally intensive and power hungry HT, including CIS, asynchronous interleaved sampling (AIS), phase-locked zero-crossing detection (PL-ZCD) and phase-locked peak-picking (PL-PP). We try to optimally balance the quality of the sound that can be conveyed via a set of pulse trains to the stimulation electrodes and hardware complexity of the processors. Then we will find that to realize a CI processor based on PL-PP strategy, an envelope detector (ED) is no longer required. This leads to a great reduction in hardware complexity and power consumption.

In Section A.2, the general concept behind all existing CI processors is described. Using this general concept, all the different processing strategies are examined and compared in terms of complexity and quality of the coded waveforms in Section A.3. To estimate the sound coding performance, the spike-based reconstruction technique [111] is applied to find the correlation factors of the input signal and the coded output pulse sequences. The results are discussed in Section A.4. Section A.5 presents a CMOS subthreshold switched–current (SI) peak-instant detector as a basic cell for compact PL-PP analog bionic ear processors. Finally, the conclusions are given in Section A.6.

## A.2 Speech Processing in Cochlear Implants

Figure A.1 shows a general block diagram that can be used to describe all strategies that are being considered. The processor comprises three layers of operation. On Layer-1, indicated by the white background boxes, the incoming sound is pre-emphasized by either amplifying or filtering or both before entering a bank of bandpass filters (BPFs). This mechanism is adapted from the role of the outer hair cells that map the wide range of the incoming sound pressure onto the limited dynamic range of the ear.

The BPF bank roughly mimics the basilar membrane behavior by decomposing the signal into a limited number (N) of frequency bands (channels). The signal strength of each channel will be extracted in the form of the temporal envelope (a process which roughly emulates the role of inner hair cell)

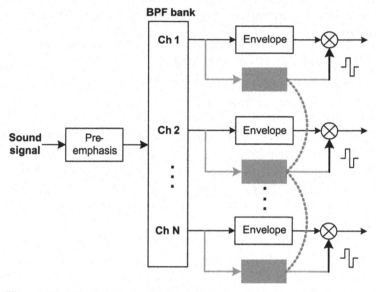

**Figure A.1** General block diagram for envelope-based speech processing.

and then modulated with the generated pulse trains to further stimulate the nerve fibers. This is common for all envelope-based processors.

The spiking pattern somewhat depends on the input frequency so that Layer-2 (indicated by the gray boxes) is introduced. Particular features (frequency, phase, TFS) of the output waveform of each channel will be detected and combined with the envelope to define the suitable stimulation pulse features. On this layer, the pulses generated from each channel are independent from each other and the stochastic spiking behavior of the auditory nerve is ignored.

Layer-3 is therefore added to include this phenomenon by somehow conditioning the features detected from different channels to create a stimulation pattern that avoids electrode interaction and preserves the relevant extracted features. Note that the attempt to convey all the features of the incoming signal to the stimulation electrode is based on the assumption that the brain can interpret this information but in practice there are factors that deviate from what really happens along the auditory pathway. So the number of layers (system complexity) does not guarantee the quality of perception in real patients [112] but serves only as a first order estimation.

## A.3 Review and Comparison of Existing Processing Strategies

### A.3.1 Continuous Interleaved Sampling

From the default setting of several CI models [105], it can be said that CIS is the most successful strategy to date. CIS employes the $1^{st}$ layer of operation. There is some evidence that the quality of speech perception obtained from a CIS processor strongly depends on the precision of the extracted envelopes [108, 113]. Accordingly, an attempt to replace the simple ED comprising a rectifier and a lowpass filter (LPF) by a HT based ED is of interest. This issue needs to be carefully considered for an analog processor since in order to perform the HT, a high complexity of its constituting electronic circuitry is unavoidable [110].

Figure A.2(a) shows a fraction of the speech signal from the word 'die' after $4^{th}$-order butterworth BP filtering with a center frequency of 150 Hz. The envelopes are extracted by a simple ED with 200 Hz LPF cutoff frequency (the dashed line of Figures A.2(b)–A.2(c) and A.2(e)) and by the HT-ED (dotted line of Figures A.2(b)–A.2(e)). The positive pulse train generated within the CIS processor is represented by the blue line in Figure A.2(b). In this case, we can clearly see that the accuracy of the amplitude of the pulses highly depends on the accuracy of the ED. Also, it is hardly possible that the brain can recognize frequency, phase and TFS from the fixed timing interval of the pulse train.

### A.3.2 Zero-Crossing Detection

In this strategy, the $2^{nd}$ layer is put on top of the $1^{st}$ layer to introduce a phase locking amplitude modulated pulse train [114]. At the moment that the input signal crosses zero from negative to positive values, a pulse is generated and will be modulated with the momentary value of the envelope at that moment to create the stimulation pulse train. It can be seen from the blue line of Figure A.2(c) that the pulse magnitudes are also defined by the quality of the ED but the real-time period of the fundamental frequency ($f_0$) can be roughly encoded. This processor thus requires high precision zero crossing instant and high accuracy envelope detectors.

### A.3.3 Peak-Picking Technique

This strategy also contains two layers of operation (1 and 2). But instead of detecting the zero-crossing moments to create the phase-locked pulse train,

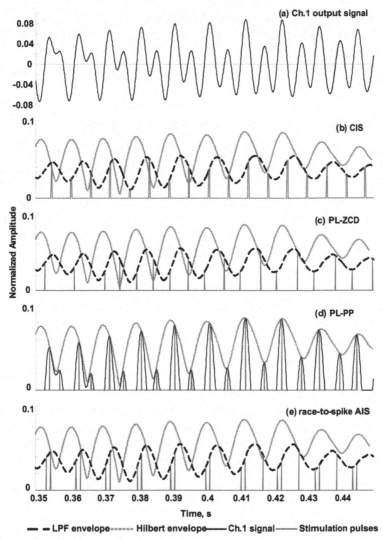

**Figure A.2**  Waveforms obtained from different strategies.

the occurrences of peaks in the input signal are detected [115]. There are two main features different from the PL-ZCD. First, the number of peaks detected is higher than the number of zero crossing moments which can be seen from the blue lines in Figures A.2(c) and A.2(d) during $0.35\text{s} < t < 0.37\text{s}$. This implies that more instantaneous frequency information other than $f_0$ can be conveyed to the stimulation electrodes. Second, as we can see from the peaks that always touch the Hilbert envelope, the BPF output signal and the detected peaking moments can be used to generate the stimulation pulses directly without the need for an ED. For this reason, the precision of the stimulation pulse amplitudes is relayed onto the precision of a peak-instant detector.

### A.3.4 Race-to-Spike Asynchronous Interleaved Sampling

In this case the $3^{rd}$ layer is introduced. It has been proposed in [111] that to achieve the stochastic stimulation behavior, the gray boxes of Figure A.1 are replaced by half-wave rectifier circuits. Then, at particular repetitive time instants, the amplitudes from all channels will be sent to a winner-take-all (WTA) network letting only the strongest amplitude pass to enable the pulse generator. To avoid successive stimulation within one channel that violates the bio-realism spiking behavior [116], additional circuit blocks are inserted to create an inhibition. At the moment that a stimulation pulse is being generated, there will be a signal created and applied to inhibit the signal from the half-wave rectifier (within the channel that is being stimulated), so that it will not be processed by the WTA network at the next time step. Even if the signal strength of that channel is highest, it will be ignored. Within this processor, the amplitude of each pulse is still specified by the ED of each channel but the location of the stimulated electrode is defined by the strength of the signal at that moment. The pulse waveform obtained from this processor is shown by the blue line in Figure A.2(e). To some extent, encoding sound using this strategy can emulate the random spiking behavior of the normal auditory nerve fiber and the perception of music is expected. It is unfortunate that the system is very complicated, requiring two more additional circuit blocks.

## A.4 System Simulations

MATLAB was used to simulate all encoding strategies. Three kinds of sounds were picked up for the simulation with a sampling frequency of 11,025 Hz including the word *'die'*, the sentence *'the discrete Fourier transform of a real value signal is conjugate symmetric'* and the sung phrase *'Hallelujah'*. An 8-channel $4^{th}$-order Butterworth BPF bank is used for all strategies with center frequencies ranging from 150 Hz to 4,000 Hz arranged according to the ERB scheme [117].

Each envelope detector is formed by a full-wave rectifier followed by a 4th-order Chebyshev LPF with 200 Hz cutoff frequency. The envelope detector is applied for all processors except PL-PP since this strategy does not need one. The stimulation pulses obtained from channels 1 to 8 of all strategies (shown in Figure A.2 only for the 1st channel) are collected for reconstruction using the spike-based technique [111] (in this context, spike refers to pulse signal). This technique has its foundation in prior neurophysiology work showing that the original analog waveform can be accurately reconstructed from a spiking waveform [118]. We therefore use this technique for the signal reconstruction.

Figure A.3 shows a block diagram of this reconstruction technique. The stimulation pulses of each channel are multiplied by uniformly distributed random noise before injecting into the BPF with the same center frequency as in the processor. The resulting signals from all channels were added to produce the output sound.

To exemplify the reconstructed waveforms, Figures A.4 and A.5 show the reconstructed sounds of the word '*die*' from the CIS and PL-PP strategies, respectively. The original and reconstructed signals are represented by the green and blue lines, respectively. Roughly, it is visible that the reconstructed signal from PL-PP is closer to its origin than that of CIS.

The correlation factor ($r$) between the original signal and the reconstructed signal was used to estimate the quality of the signals encoded from different strategies. The correlation coefficient is computed from the following equation where $X_i$, $Y_i$, $\bar{X}$ and $\bar{Y}$ are the original, the reconstructed signals, the mean value of $X_i$ and the mean value of $Y_i$, respectively.

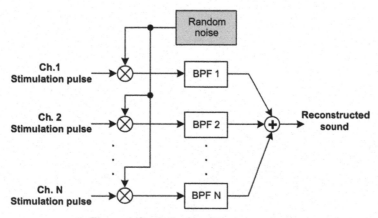

**Figure A.3**   Spike-based reconstruction.

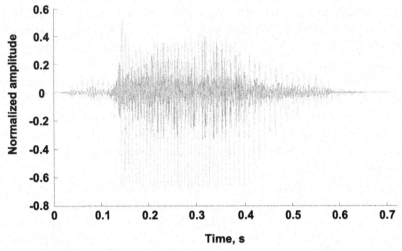

**Figure A.4**    Reconstructed waveform of the word *'die'* from the CIS.

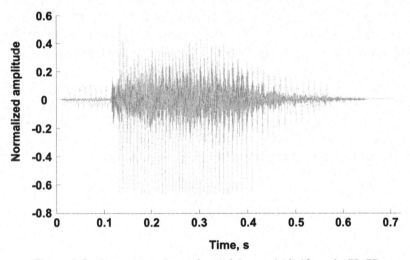

**Figure A.5**    Reconstructed waveform of the word *'die'* from the PL-PP.

$$r = \frac{\sum\limits_{i=0}^{n} (X_i - \bar{X})(Y_i - \bar{Y})}{\sqrt{\sum\limits_{i=0}^{n} (X_i - \bar{X})^2} \sqrt{\sum\limits_{i=0}^{n} (Y_i - \bar{Y})^2}}, \tag{A.1}$$

Table A.1   Correlation for different strategies

| Strategy | 'die' | Sentence | 'hallelujah' |
|---|---|---|---|
| CIS | 0.11 | 0.05 | 0.02 |
| PL-ZCD | 0.36 | 0.14 | 0.50 |
| PL-PP | 0.47 | 0.25 | 0.59 |
| race-to-spike AIS | 0.49 | 0.38 | 0.52 |

The correlation coefficient varies in the range of −1 to 1, where 1 indicates a perfect correlation, −1 shows the inversely perfect correlation and 0 represents non correlation. The resulting correlation coefficients obtained from different strategies are shown in Table A.1. It is clear that CIS performs worst of all. Besides, within the results from CIS, the values of $r$ depend on the complexity of the original sounds. The highest value of $r = 0.11$ is from the simple word (single tone) and the lowest $r = 0.02$ is from the song which contains several tones that CIS could not capture. Among the PL processors, as expected from the coding mechanism that conveys more instantaneous information without loss from the non-ideality of the ED (see Figure A.2(d)), the PL-PP provides a better value of $r$ than PL-ZCD for all cases. The race-to-spike AIS processor gives the best values of $r$ for less complicated sounds (word and sentence) but for multi-tone sounds (song), the highest value of $r$ is given by the PL-PP processor.

## A.5  Discrete-Time Peak-Instant Detector

The comparative study in the previous section suggests that the PL-PP strategy provides a compact solution to partially convey the TFS suitable for an ultra-low-power analog BE processor. This section presents a possible compact realization of a circuit called 'peak-instant detector (PID)' using the SI technique outlined in Part I.

Figure A.6 shows a block diagram of the PL-PP speech processing strategy. It comprises four main elements; a pre-emphasis block, a bandpass filter (BPF) bank, a number of PIDs, and modulators. The incoming sound is pre-emphasized by non-linearly amplifying it before entering the bank of BPFs. This mechanism is adapted from the role of the outer hair cells that map the wide range of the incoming sound pressure onto the limited dynamic range of the ear. The BPF bank roughly mimics the basilar membrane behavior by decomposing the signal into a limited number (N) of frequency bands (channels). The peaks of the signal coming from BPF of each channel will be extracted by the PID enabling the modulator to perform multiplication of

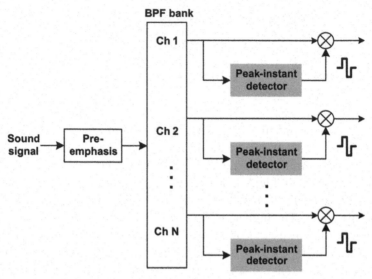

**Figure A.6**   PL-PP speech processing block diagram.

a rectangular pulse signal and the peak amplitude. This results in a set of rectangular stimulation pulses of which the amplitude is defined by the peak of the signal of each channel and the pulse frequency change according to the speech signal of each channel.

Note that there is a possibility that the resulting stimulation pulses from different channels can appear at the same time introducing simultaneous inter-actions between stimulation electrodes which degrades the hearing perception. This undesired mechanism still needs to be eliminated.

Figure A.7 shows a fraction of the speech signal from the word *'die'* after $4^{th}$-order butter-worth BP filtering with a center frequency of 150 Hz (Ch. 1 of the block diagram in Figure A.6). The signal obtained from Ch. 1 of the BPF bank is represented by the gray line. The HT (which is not part of the PL-PP strategy) is applied to the gray line signal and the envelope is extracted and shown by the dotted line. The resulting pulses taken from the output of the modulator are represented by the solid line. As we can see from the peaks that always touch the Hilbert envelope, the detected peaking moments of the BPF output signal can be used to generate the stimulation pulses directly without the need for a very precise envelope detector. The precision requirement of the stimulation pulse amplitudes is relayed onto the precision of a PID instead.

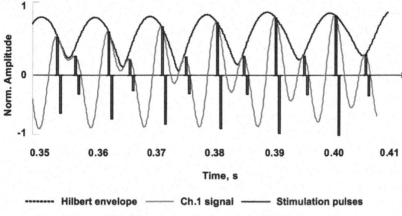

Figure A.7   Waveforms obtained from PL-PP strategies.

## A.5.1 Switched-Current PID Concept

The basic idea of the proposed PID is shown by the block diagram in
Figure A.8(a) [119, 120]. Input signal $x_t$ is split into two signal paths. First, it
goes to the sample and hold amplifier, SHA, (with a unity gain) generating a
half delayed signal, $x_{td}$. Second, it goes to the summing node. The sample and
hold time period are controlled by a clock signal with 50% duty cycle indicated
as the middle trace of Figures A.8(b) and A.8(c). The subtracted result of $x_t$

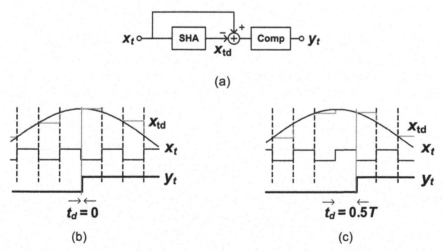

**Figure A.8**   SI peak detection concept: (a) Operational block diagram. (b) Best case and
(c) worst case detections.

and $x_{td}$ at the holding period will change its polarity when $x_{td}$ reaches its maximum and minimum values. For this reason, the comparator can decide on its logical output $y_t$ according to its input sign reversing moment. Since the incoming signal is random, this concept gives us two extreme cases of delay time (assuming the comparator is ideal).

Firstly, the minimum delay time $(t_d)$ occurs when the falling edge of the clock signal is located exactly at the peak of $x_t$ (See Figure A.8(b)). In this case, the detected $y_t$ will not be delayed.

Secondly, the maximum delay time occurs when the rising edge of the clock is located exactly at the peak (See Figure A.8(c)). In this case, $t_d$ becomes a half period of the clock signal. In practice, the charge injection error and output noise of the SHA and minimum detectable input signal and delay time of the comparator introduce additional errors. In circuit level design, the architecture of the SHA needs to be insensitive to the switch charge injection error and the resolution and speed of the comparator need to be sufficiently good to decide on its logical output within a half clock period.

### A.5.2 Switched-Current PID Circuit

Figure A.9 shows a macro-model of the proposed PID. It comprises a fully differential SI-SHA and a voltage comparator. The PID is controlled by two non-overlapping clock phases $S_1$ (sampling) and $S_2$ (holding). When the set of switches $S_1$ turns on, the differential input current will be converted into voltages across $C_H$. In the next phase, $S_1$ turns off and the set of switches $S_2$ turns on. Both identical $C_H$'s will memorize the voltages across them producing a constant differential current via transconductor $G_m$. The memorized current will be compared with the input current and converted into a differential voltage at the input nodes of the comparator. The comparator will make a decision within this phase and generate binary output voltage $V_{out}$. Due to the large loop gain provided by voltage amplifier $A_v$, the voltages across switches $S_1$ are forced to be fixed at the input common mode level. This leads to signal-independent charge injection errors after the (practical MOS) switches are turned off and thus will be cancelled out by the differential operation at the input of the comparator [3].

The sub-circuits used to realize all active elements in Figure A.9 are shown in Figure A.10. $A_v$ is formed by the circuit of Figure A.10(a) and its output common-mode level $V_{C2}$ is controlled by the common-mode feedback (CMFB) circuit depicted in Figure A.10(b). $G_m$ is realized by the circuit

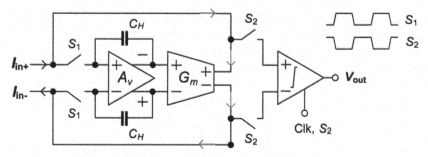

**Figure A.9** Macromodel of the SI peak-instant detector.

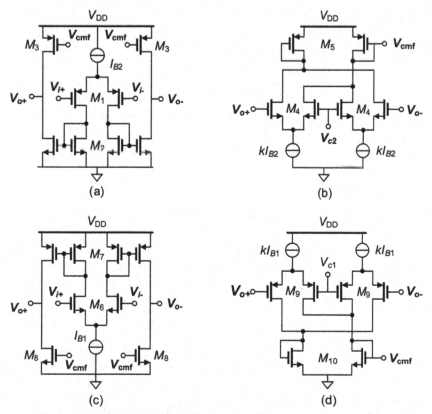

**Figure A.10** Sub-circuits of the SHA. (a) Transconductor $G_m$, (b) CMFB circuit for $G_m$, (c) voltage amplifier $A_v$ and (d) CMFB circuit for $A_v$.

in Figure A.10(c) and its output common-mode level $V_{C1}$ is controlled by the CMFB circuit in Figure A.10(d). Stability of the feedback loop can be maintained by setting a fixed ratio of bias current $I_{B2}$ and $I_{B1}$ and a value of $C_H$ that needs to be bigger than the parasitic capacitances present at the input nodes of $A_v$ and those of $G_m$ when the loop is closed [116]. This condition can be satisfied by setting $I_{B2} = 2.2\,I_{B1} = 220$ nA and realizing $C_H$ by NMOS capacitors biased in strong inversion region. We thus set supply voltage $V_{DD} = 1.2$ V, common-mode voltages $V_{C1} = 1$ V and $V_{C2} = 0.2$ V. All the transistors in the entire circuit are working in weak inversion as their parasitic capacitances are smaller compared to those of MOS transistors in strong inversion region for the same device size.

The comparator is realized by the circuit shown in Figure A.11. It is composed of a differential input stage followed by a chain of CMOS inverters to enhance the overall gain. Input parasitic capacitors $C_{ip}$ and $C_{in}$ are employed to memorize the input voltage throughout the whole sampling period (switches $S_2$ are turned-off). It is worth to mentioning that the comparator employed here is just a simple high gain open-loop amplifier which does not provide high speed and high sensitivity. Better results can be expected from using a more sophisticated comparator circuit, if needed.

### A.5.3 Circuit Simulation

The PID circuit has been designed using AMIS 0.35-$\mu$m CMOS process technology parameters. The bias current of the comparator is set to $I_{B3}=$ 50 nA. The total bias current becomes 722 nA (excluding that of the bias generator circuit). This results in a static power consumption of 866.4 nW. Dimensions of the MOS transistors used are listed in Table. A.2. Large area

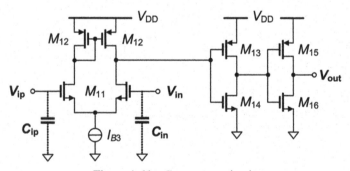

**Figure A.11**    Comparator circuit.

transistors are used to alleviate the mismatch problem of MOS transistors in weak inversion.

Figure A.12 demonstrates the transient response of the proposed PID circuit in the worst case detection (see Figure A.8(c)) for a sinusoidal differential input current with an amplitude of 80 nA, a 5 kHz frequency and a 100 kS/s sampling frequency. The input current and the holding current are shown on the top by the dotted and solid lines, respectively. There are large transient glitches appearing at the beginning of the holding phases but they do not affect the correct circuit operation. In the middle graph, we can see that the voltages at the input nodes of the comparator are swinging up and down crossing each other within the first holding period after the peak occurred. This operation is consistent with the theory explained earlier but, as we can see from the output waveform $V_{out}$ shown in the bottom graph, the comparator produces an additional time delay. Also it can be seen that the delay time for the negative peak is slightly shorter.

In Figure A.13, it is indicated that the delay times of the PID circuit for both positive and negative peaks depend on the input amplitude. For very small input amplitudes less than 50 nA the proposed circuit gives a delay time bigger than 10 $\mu$s which is 5% of the period of the input signal. This is due to the limited resolution of the comparator. For the range of input amplitude of 50 nA to 100 nA, the delay time remains less than 5% of the input signal period. The delay time goes up again for input amplitudes larger than 100 nA. This is not because of a limitation of the comparator but of the SHA. As the internal voltage swings at the input of $G_m$ go too high, the charge injection error cannot be completely cancelled out leading to a wrong decision of the comparator.

Since the mismatch in weak inversion is worse than in strong inversion, a Monte Carlo simulation has been performed to verify the circuit operation. For the same condition of the transient response shown in Figure A.12, with 300 runs, it gives a mean values ($\langle x \rangle$) and a standard deviation ($\sigma$) of 8.8 $\mu$s

**Table A.2**  Transistor dimensions

| MOSFET | W [$\mu$m] | L [$\mu$m] |
|---|---|---|
| $M_1$, $M_4$, $M_6$, $M_9$ | 24 | 6 |
| $M_2$, $M_3$, $M_5$, $M_7$, $M_8$, $M_{10}$ | 3 | 3 |
| NMOS $M_H$ ($C_H$) | 11 | 11 |
| $M_{11}$ | 6 | 3 |
| $M_{12}$ | 3 | 3 |
| $M_{13}$, $M_{14}$, $M_{15}$, $M_{16}$ | 0.5 | 0.35 |

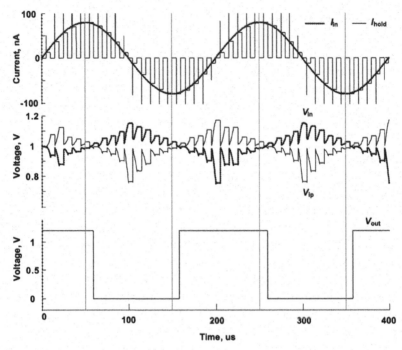

**Figure A.12**   Signal swings within the proposed PID circuit.

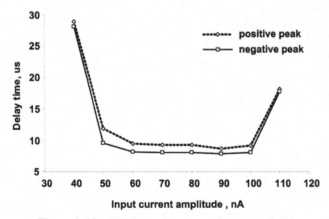

**Figure A.13**   PID delay time versus input amplitude.

and 2.16 $\mu$s for the positive peak and $\langle x \rangle$ = 8.2 $\mu$s and $\sigma$ = 2.4 $\mu$s for the negative one. These numbers indicate that the delays spread around 5% of the input signal's period.

## A.6 Discussion and Conclusions

The system complexity and quality of the reconstructed signals from different signal processing strategies have been investigated and compared. Targeting the design of a fully implantable analog bionic ear with an ability of tone recognition, the PL-PP strategy provides us the best solution, both in terms of compactness and correlation factors. Since the information of frequency, phase and TFS cannot be conveyed to the stimulation electrodes by CIS, it is really hard to believe that the brain can recognize any tone without proper input information. CIS is therefore removed from our consideration.

One may object that the values of the correlation factor cannot 100% guarantee the quality of hearing perception in real BE recipients. We are still optimistic that the brain can interpret multi-tone sounds from the fast varying information conveyed to the stimulation electrodes by the rest of the strategies and that, after long term training, the patients would be able to recognize tonal languages and melodies.

Regarding the hardware implementation, an ultra-low-power PID designed for a PL-PP BE processor has been also presented. The instants detected are delayed within less than one clock period even if the transistors' mismatch is taken into account. Either the rising or the falling edges of the output signal together with the input signal amplitude are expected to be used as control parameters in a stimulator for cochlear epical electrodes which operates in the frequency range of 300 Hz–5 kHz.

Combining the aforementioned ability of extracting TFS with the feasibility of building ultra-low-power analog hardware, the PL-PP has proven itself as a suitable processor for a contemporary fully implantable analog bionic ear.

# Appendix B

## Harmonic Distortion Calculation
## for Chapter 3

At the end of the sampling phase, the nonlinearity of, and the mismatch between capacitors $C_H$, the mismatch between switches $S_2$ and an insufficient LG lead to incomplete switching error compensation. This residual error can be modeled as an input offset voltage $V_{offSW}$ to $G_{m2}$, which appears during the hold phase only and equals

$$V_{offSW} = V_{CFT+} - V_{CFT-} \tag{B.1}$$

Where $V_{CFT+}$ and $V_{CFT-}$ are error voltages induced by charge injection and clock-feedthrough effects of the MOS switches appearing on the non-inverting and inverting terminals of $G_{m2}$, respectively. Effects of $V_{offSW}$ will be shown for the class-A and class-AB circuits, respectively, in the following paragraph.

During the hold phase $V_{offSW}$ is added to the differential input voltage, $V_{id}$, leading to

$$I_{od} = I_{B1} \tanh\left(\frac{V_{id} + V_{offSW}}{2n_p V_T}\right), \tag{B.2}$$

and

$$I_{od} = 2I_{B2} \sinh\left(\frac{V_{id} + V_{offSW}}{n_p V_T}\right). \tag{B.3}$$

### B.1 Class A CSH Circuit

Considering (B.2) we can find that

$$I_{od} = I_{B1} \tanh\left(\frac{V_{id} + V_{offSW}}{2n_p V_T}\right)$$

155

$$= I_{B1} \frac{\tanh\left(\frac{V_{id}}{2n_pV_T}\right) + \tanh\left(\frac{V_{offsw}}{2n_pV_T}\right)}{1 + \tanh\left(\frac{V_{id}}{2n_pV_T}\right) \cdot \tanh\left(\frac{V_{offsw}}{2n_pV_T}\right)} = I_{B1}\left(\frac{A+B}{1+A\cdot B}\right), \quad (B.4)$$

where $A = \tanh\left(\frac{V_{id}}{2n_pV_T}\right) = \frac{I_{id}}{I_{B1}}$ and $B = \tanh\left(\frac{V_{offsw}}{2n_pV_T}\right) = \frac{I_E}{I_{B1}}$.

Applying a Taylor series expansion to (B.4), we have

$$I_{od} \cong I_{B1}\left[B + A(1-B^2) + A^2(B^3-B) + A^3(B^2-B^4) + A^4(B^5-B^3) + \cdots\right]. \quad (B.5)$$

As $A = \frac{I_{id}}{I_{B1}}$ and $B = \frac{I_E}{I_{B1}}$, (B.5), can be arranged to

$$\frac{I_{od}}{I_{B1}} \cong a_0 + a_1\frac{I_{id}}{I_{B1}} + a_2\left(\frac{I_{id}}{I_{B1}}\right)^2 + a_3\left(\frac{I_{id}}{I_{B1}}\right)^3 + a_4\left(\frac{I_{id}}{I_{B1}}\right)^4 + \cdots, \quad (B.6)$$

where $a_0 = \frac{I_E}{I_{B1}}$, $a_1 = \left(1 - \left(\frac{I_E}{I_{B1}}\right)^2\right)$, $a_2 = \left(\left(\frac{I_E}{I_{B1}}\right)^3 - \frac{I_E}{I_{B1}}\right)$,

$a_3 = \left(\left(\frac{I_E}{I_{B1}}\right)^2 - \left(\frac{I_E}{I_{B1}}\right)^4\right)$ and $a_4 = \left(\left(\frac{I_E}{I_{B1}}\right)^5 - \left(\frac{I_E}{I_{B1}}\right)^3\right)$.

Substituting $I_{id} = \hat{I}_{id}\cos(\omega t)$ into (B.6), we found that

$$\frac{I_{od}}{I_{B1}} \cong a_0 + a_1\frac{\hat{I}_{id}}{I_{B1}}\cos(\omega t)$$

$$+\frac{a_2}{2}\left(\frac{\hat{I}_{id}}{I_{B1}}\right)^2(1+\cos(2\omega t))$$

$$+\frac{a_3}{4}\left(\frac{\hat{I}_{id}}{I_{B1}}\right)^3(3\cos(\omega t)+\cos(3\omega t))$$

$$+\frac{a_4}{4}\left(\frac{\hat{I}_{id}}{I_{B1}}\right)^4\left(\frac{3}{2}+2\cos(2\omega t)+\cos(4\omega t)\right) + \cdots \quad (B.7)$$

The second, third and fourth harmonic distortions (HD$_{2A}$, HD$_{3A}$ and HD$_{4A}$) can be extracted from (B.7) as

$$HD_{2A} = \frac{\frac{a_2}{2}\left(\frac{\hat{I}_{id}}{I_{B1}}\right)^2 + \frac{a_4}{2}\left(\frac{\hat{I}_{id}}{I_{B1}}\right)^4}{\left(a_1\frac{\hat{I}_{id}}{I_{B1}} + \frac{3a_3}{4}\left(\frac{\hat{I}_{id}}{I_{B1}}\right)^3\right)} \cong \frac{a_2\left(\frac{\hat{I}_{id}}{I_{B1}}\right)}{2a_1} \cong \frac{1}{2}\frac{a_2}{a_1}MI_1,$$

$$\mathrm{HD}_{3A} = \frac{\frac{a_3}{4}\left(\frac{\hat{I}_{id}}{I_{B1}}\right)^3}{\left(a_1\frac{\hat{I}_{id}}{I_{B1}} + \frac{3a_3}{4}\left(\frac{\hat{I}_{id}}{I_{B1}}\right)^3\right)} \cong \frac{a_3\left(\frac{\hat{I}_{id}}{I_{B1}}\right)^2}{4a_1} \cong \frac{1}{4}\frac{a_3}{a_1}(MI_1),$$

and

$$\mathrm{HD}_{4A} = \frac{\frac{a_4}{4}\left(\frac{\hat{I}_{id}}{I_{B1}}\right)^4}{\left(a_1\frac{\hat{I}_{id}}{I_{B1}} + \frac{3a_3}{4}\left(\frac{\hat{I}_{id}}{I_{B1}}\right)^3\right)} \cong \frac{a_4\left(\frac{\hat{I}_{id}}{I_{B1}}\right)^3}{4a_1} \cong \frac{1}{4}\frac{a_4}{a_1}(MI_1)^3,$$

As $\frac{I_{id}}{I_{B1}} = \tanh\left(\frac{V_{id}}{n_p V_T}\right)$ and $\frac{I_E}{I_{B1}} = \tanh\left(\frac{V_{offSW}}{n_p V_T}\right)$, it can also be found that

$$\mathrm{HD}_{2A} = \frac{\tanh\left(\frac{V_{offSW}}{2n_p V_T}\right)\left(\tanh^2\left(\frac{V_{offSW}}{2n_p V_T}\right) - 1\right)}{1 - \tanh^2\left(\frac{V_{offSW}}{2n_p V_T}\right)}(MI_1),$$

$$\mathrm{HD}_{3A} = \tanh^2\left(\frac{V_{offSW}}{2n_p V_T}\right)(MI_1)^2,$$

and

$$\mathrm{HD}_{4A} = \frac{\tanh^4\left(\frac{V_{offSW}}{2n_p V_T}\right)\left(\tanh\left(\frac{V_{offSW}}{2n_p V_T}\right) - 1\right)}{1 - \tanh^2\left(\frac{V_{offSW}}{2n_p V_T}\right)}(MI_1)^3,$$

where $MI_1 = \hat{I}_{id}/I_{B1}$ is the modulation index of the class-A transconductor and $\hat{I}_{id}$ represents the amplitude of the sinusoidal input current $I_{id}$.

## B.2 Class-AB CSH Circuit

Considering (B.2), it can be expanded that

$$I_{od} = 2I_{B2}\left[\sinh\left(\frac{V_{id}}{n_p V_T}\right)\cdot \cosh\left(\frac{V_{offSW}}{n_p V_T}\right) + \cosh\left(\frac{V_{id}}{n_p V_T}\right)\right.$$
$$\left.\cdot \sinh\left(\frac{V_{offSW}}{n_p V_T}\right)\right], \tag{B.8}$$

Since $\sinh\left(\frac{V_{id}}{n_p V_T}\right) = \frac{I_{id}}{2I_{B2}}$ and we define that $\sinh\left(\frac{V_{offSW}}{n_p V_T}\right) = \frac{I_E}{2I_{B2}}$, (B.8) can be re-arranged to

$$I_{od} = I_{id} \cosh\left(\frac{V_{offSW}}{n_p V_T}\right) + I_E \cosh\left(\frac{V_{id}}{n_p V_T}\right). \qquad \text{(B.9)}$$

As $V_{id} = n_p V_T \, \mathrm{a}\sinh\left(\frac{I_{id}}{2I_{B2}}\right)$, (B.9) becomes

$$\frac{I_{od}}{I_{B2}} = \frac{I_{id}}{I_{B2}} \cosh\left(\frac{V_{offSW}}{n_p V_T}\right) + \frac{I_E}{I_{B2}} \cosh\left(\mathrm{a}\sinh\left(\frac{I_{id}}{2I_{B2}}\right)\right). \qquad \text{(B.10)}$$

Applying a Taylor series expansion to (B.10), we have

$$\frac{I_{od}}{I_{B2}} = b_0 + b_1\left(\frac{I_{id}}{I_{B2}}\right) + b_2\left(\frac{I_{id}}{I_{B2}}\right)^2 + b_3\left(\frac{I_{id}}{I_{B2}}\right)^3 + b_4\left(\frac{I_{id}}{I_{B2}}\right)^4, \qquad \text{(B.11)}$$

where $b_0 = \frac{I_E}{I_{B2}} = 2\sinh\left(\frac{V_{offSW}}{n_p V_T}\right)$, $b_1 = \cosh\left(\frac{V_{offSW}}{n_p V_T}\right)$, $b_2 = \frac{1}{8}\left(\frac{I_E}{I_{B2}}\right) = \frac{1}{4}\sinh\left(\frac{V_{offSW}}{n_p V_T}\right)$, $b_3 = 0$ and $b_4 = -\frac{1}{128}\left(\frac{I_E}{I_{B2}}\right) = -\frac{1}{64}\sinh\left(\frac{V_{offSW}}{n_p V_T}\right)$.

Substituting $I_{id} = \hat{I}_{id}\cos(\omega t)$ into (B.11), we found that

$$\mathrm{HD}_{2AB} = \frac{\frac{a_2}{2}\left(\frac{\hat{I}_{id}}{I_{B2}}\right)^2 + \frac{a_4}{2}\left(\frac{\hat{I}_{id}}{I_{B2}}\right)^4}{\left(a_1\frac{\hat{I}_{id}}{I_{B2}}\right)}$$

$$= \frac{1}{8}\tanh\left(\frac{V_{offSW}}{n_p V_T}\right)\mathrm{MI}_2\left(1 - \frac{1}{16}(\mathrm{MI}_2)^2\right),$$

and

$$\mathrm{HD}_{4AB} = \frac{\frac{a_4}{4}\left(\frac{\hat{I}_{id}}{I_{B2}}\right)^4}{\left(a_1\frac{\hat{I}_{id}}{I_{B2}}\right)} = \frac{1}{256}\tanh\left(\frac{V_{offSW}}{n_p V_T}\right)(\mathrm{MI}_2)^3,$$

where $\mathrm{MI}_2 = \hat{I}_{id}/I_{B2}$ is the modulation index of the class-AB transconductor.

# References

[1] C. Sawigun and W. Serdijn, "Ultra-low-power, class-AB, CMOS four-quadrant current multiplier," *Electronics Letters*, vol. 45, no. 10, pp. 483–484, 2009.

[2] C. Sawigun and W. A. Serdijn, "A nano-power class-AB current multiplier for energy-based action potential detector," in *Proc. European Conference on Circuit Theory and Design (ECCTD 2009)*, pp. 417–420, Aug. 2009.

[3] C. Sawigun and W. A. Serdijn, "A 24-nW, 0.65-V, 74-dB SNDR, 83-dB DR, class-AB current-mode sample and hold circuit," in *Proc. 2010 IEEE International Symposium on Circuits and Systems (ISCAS)*, pp. 3132–3135, June 2010.

[4] C. Sawigun, W. Ngamkham, and W. A. Serdijn, "Comparison of speech processing strategies for the design of an ultra low-power analog bionic ear," in *Proc. 2010 Annual International Conference of the IEEE Engineering in Medicine and Biology Society (EMBC)*, pp. 1374–1377, Sep. 2010.

[5] C. Sawigun, W. Ngamkham, and W. A. Serdijn, "An ultra low-power peak-instant detector for a peak picking cochlear implant processor," in *Proc. 2010 IEEE Biomedical Circuits and Systems Conference (BioCAS)*, pp. 222–225, Nov. 2010.

[6] C. Sawigun and W. Serdijn, "Analysis and design of a low-voltage, low-power, high-precision, class-AB current-mode subthreshold CMOS sample and hold circuit," *IEEE Transactions on Circuits and Systems I: Regular Papers*, vol. 58, pp. 1615–1626, June 2011.

[7] C. Sawigun, W. Ngamkham, and W. Serdijn, "A 2.6-nW, 0.5-V, 52-dB-DR, $4^{th}$-order $G_m–C$ BPF: Moving closer to the FoM's fundamental limit," in *Proc. 2012 IEEE International Symposium on Circuits and Systems (ISCAS)*, pp. 656–659, May 2012.

[8] C. Sawigun and W. A. Serdijn, "A modular transconductance reduction technique for very low-frequency $G_m–C$ filters," in *Proc. 2012 IEEE*

*International Symposium on Circuits and Systems (ISCAS)*, pp. 1183–1186, May 2012.

[9] C. Sawigun, S. Hiseni, and W. A. Serdijn, "A 0.45-nW, 0.5-V, 59-dB DR, $G_m-C$ lowpass filter for portable ECG recording," in *Proc. International Conference on Biomedical Electronics and Devices*, pp. 1–4, Feb. 2012.

[10] H. Fodstad and M. Hariz, "Electricity in the treatment of nervous system disease," in *Operative Neuromodulation* (D. Sakas, B. Simpson, and E. Krames, eds.), vol. 97/1 of *Acta Neurochirurgica Supplements*, pp. 11–19, Springer Vienna, 2007.

[11] Cavallo, "A complete treatise on electricity, in theory and practice, with original experiments." http://archive.org/details/completetreatise03cav auoft, 1777.

[12] K. Wise, A. Sodagar, Y. Yao, M. Gulari, G. Perlin, and K. Najafi, "Microelectrodes, microelectronics, and implantable neural microsystems," *Proceedings of the IEEE*, vol. 96, pp. 1184 –1202, july 2008.

[13] O. Schanne, M. Lavallee, R. Laprade, and S. Gagne, "Electrical properties of glass microelectrodes," *Proceedings of the IEEE*, vol. 56, pp. 1072–1082, june 1968.

[14] D. Robinson, "The electrical properties of metal microelectrodes," *Proceedings of the IEEE*, vol. 56, pp. 1065–1071, june 1968.

[15] M. Ortmanns, A. Rocke, M. Gehrke, and H.-J. Tiedtke, "A 232-channel epiretinal stimulator ASIC," *IEEE Journal of Solid-State Circuits*, vol. 42, pp. 2946–2959, dec. 2007.

[16] B. S. Wilson and M. F. Dorman, "Cochlear implants: A remarkable past and a brilliant future," *Hearing Research*, vol. 242, pp. 3–21, 2008.

[17] S. Breit, J. B. Schulz, and A.-L. Benabid, "Deep brain stimulation," *Cell and Tissue Research*, vol. 318, pp. 275–288, 2004.

[18] P. R. Troyk and N. d. N. Donaldson, "Implantable FES stimulation systems: What is needed?," *Neuromodulation: Technology at the Neural Interface*, vol. 4, no. 4, pp. 196–204, 2001.

[19] D. Kipke, W. Shain, G. Buzsáki, E. Fetz, J. Henderson, J. Hetke, and G. Schalk, "Advanced neurotechnologies for chronic neural interfaces: New horizons and clinical opportunities," *The Journal of Neuroscience*, vol. 28, no. 46, pp. 11830–11838, 2008.

[20] X.-F. Teng, Y.-T. Zhang, C. Poon, and P. Bonato, "Wearable medical systems for p-health," *IEEE Reviews in Biomedical Engineering*, vol. 1, pp. 62–74, 2008.

[21] W. H. Organization, "Preventing chronic diseases: A vital investment.," tech. rep., WHO Global Report, 2005.

[22] A. Lymberis and A. Dittmar, "Advanced wearable health systems and applications-research and development efforts in the European union," *IEEE Engineering in Medicine and Biology Magazine*, vol. 26, no. 3, pp. 29–33, 2007.

[23] R. Harrison, "The design of integrated circuits to observe brain activity," *Proceedings of the IEEE*, vol. 96, no. 7, pp. 1203–1216, 2008.

[24] R. Harrison, R. Kier, C. Chestek, V. Gilja, P. Nuyujukian, S. Ryu, B. Greger, F. Solzbacher, and K. Shenoy, "Wireless neural recording with single low-power integrated circuit," *IEEE Transactions on Neural Systems and Rehabilitation Engineering*, vol. 17, pp. 322–329, Aug. 2009.

[25] C. C. Enz, F. Krummenacher, and E. A. Vittoz, "An analytical MOS transistor model valid in all regions of operation and dedicated to low-voltage and low-current applications," *Analog Integrated Circuits and Signal Processing*, vol. 8, no. 1, pp. 83–114, 1995.

[26] P. Kinget, "Device mismatch and tradeoffs in the design of analog circuits," *IEEE Journal of Solid-State Circuits*, vol. 40, no. 6, pp. 1212–1224, 2005.

[27] C. Enz and G. Temes, "Circuit techniques for reducing the effects of op-amp imperfections: autozeroing, correlated double sampling, and chopper stabilization," *Proceedings of the IEEE*, vol. 84, no. 11, pp. 1584–1614, 1996.

[28] C. Toumazou and J. Hughes, *Switched-currents: An analogue technique for digital technology*, vol. 5, IEE, 1993.

[29] P. M. Sinn and G. W. Roberts, "A comparison of first and second-generation switched-current cells," in *Proc. 1994 IEEE International Symposium on Circuits and Systems(ISCAS'94)*, vol. 5, pp. 301–304.

[30] G. Wegmann, E. Vittoz, and F. Rahali, "Charge injection in analog MOS switches," *IEEE Journal of Solid-State Circuits*, vol. 22, no. 6, pp. 1091–1097, 1987.

[31] C. Sawigun and W. Serdijn, "0.75-V micro-power SI memory cell with feedthrough error reduction," *IET Electronics Letters*, vol. 44, no. 9, pp. 561–562, 2008.

[32] J. B. Hughes and K. W. Moulding, "S2I: A switched-current technique for high performance," *IEE Electronics Letters*, vol. 29, no. 16, pp. 1400–1401, 1993.

[33] D. Nairn, "A high-linearity sampling technique for switched-current circuits," *IEEE Transactions on Circuits and Systems II: Analog and Digital Signal Processing*, vol. 43, no. 1, pp. 49–52, 1996.

[34] E. Sánchez-Sinencio and J. Silva-Martinez, "CMOS transconductance amplifiers, architectures and active filters: a tutorial," *IEE proceedings-circuits, devices and systems*, vol. 147, no. 1, pp. 3–12, 2000.

[35] S. D'Amico, M. Conta, and A. Baschirotto, "A 4.1-mW 10-MHz fourth-order source-follower-based continuous-time filter with 79-dB DR," *IEEE Journal of Solid-State Circuits*, vol. 41, no. 12, pp. 2713–2719, 2006.

[36] T.-T. Zhang, P.-I. Mak, M.-I. Vai, P.-U. Mak, M.-K. Law, S.-H. Pun, F. Wan, and R. P. Martins, "15-nW biopotential LPFs in 0.35-CMOS using subthreshold-source-follower biquads with and without gain compensation," *IEEE Transactions on Biomedical Circuits and Systems*, vol. 7, pp. 690–702, 2013.

[37] B. Gilbert, "Translinear circuits: A proposed classification," *IEE Electronics letters*, vol. 11, no. 1, pp. 14–16, 1975.

[38] E. Seevinck and R. J. Wiegerink, "Generalized translinear circuit principle," *IEEE Journal of Solid-State Circuits*, vol. 26, no. 8, pp. 1098–1102, 1991.

[39] Y. Taur, "The incredible shrinking transistor," *IEEE Spectrum*, vol. 36, pp. 25–29, july 1999.

[40] E. Vittoz and J. Fellrath, "CMOS analog integrated circuits based on weak inversion operations," *IEEE Journal of Solid-State Circuits*, vol. 12, no. 3, pp. 224–231, 1977.

[41] A. Worapishet and J. B. Hughes, "Performance enhancement of switched-current technique using subthreshold mos operation," *IEEE Transactions on Circuits and Systems I: Regular Papers*, vol. 55, no. 11, pp. 3582–3592, 2008.

[42] A. Annema, B. Nauta, R. van Langevelde, and H. Tuinhout, "Analog circuits in ultra-deep-submicron CMOS," *IEEE Journal of Solid-State Circuits*, vol. 40, no. 1, pp. 132–143, 2005.

[43] C. Toumazou, J. Hughes, and D. Pattullo, "Regulated cascode switched-current memory cell," *IEE Electronics Letters*, vol. 26, no. 5, pp. 303–305, 1990.

[44] D. Robertson, P. Real, and C. Mangelsdorf, "A wideband 10-bit, 20 Msps pipelined ADC using current-mode signals," in *37$^{th}$ IEEE International Solid-State Circuits Conference, Digest of Technical Papers*, pp. 160–161, IEEE, 1990.

[45] O. Landolt, "An analog CMOS implementation of a Kohonen network with learning capability," in *Proc. An international workshop on VLSI for neural networks and artificial intelligence*, (New York, NY, USA), pp. 25–34, Plenum Press, 1992.

[46] L. Marques, W. Serdijn, C. Galup-Montoro, and M. Schneider, "A switched-MOSFET programmable low-voltage filter," in *Proc. 15th Symposium on Integrated Circuits and Systems Design.*, pp. 254–257, IEEE, 2002.

[47] J. Martins and V. Dias, "Very low-distortion fully differential switched-current memory cell," *IEEE Transactions on Circuits and Systems II: Analog and Digital Signal Processing*, vol. 46, no. 5, pp. 640–643, 1999.

[48] G. Balachandran and P. Allen, "Switched-current circuits in digital CMOS technology with low charge-injection errors," *IEEE Journal of Solid-State Circuits*, vol. 37, no. 10, pp. 1271–1281, 2002.

[49] P. Gray, P. Hurst, S. Lewis, and R. Meyer, *Analysis and design of analog integrated circuits*, ch. 9. New York: Wiley, 4th ed., 2001.

[50] G. Palmisano and G. Palumbo, "A novel representation for two-pole feedback amplifiers," *IEEE Transactions on Education*, vol. 41, no. 3, pp. 216–218, 1998.

[51] W. Serdijn, *The design of low-voltage low-power analog integrated circuits and their applications in hearing instruments*. PhD thesis, Delft university of Technology, Delft, 1994.

[52] P. Shah and C. Toumazou, "A theoretical basis for very wide dynamic range switched-current analogue signal processing," *Analog Integrated Circuits and Signal Processing*, vol. 7, no. 3, pp. 201–213, 1995.

[53] M. Schlarmann and R. Geiger, "Relationship between amplifier settling time and pole-zero placements for second-order systems," in *Proc. 43rd IEEE Midwest Symposium on Circuits and Systems*, vol. 1, pp. 54–59 vol.1, 2000.

[54] O. Oliaei and P. Loumeau, "Glitch reduction in second-generation SI circuits," *IEE Electronics Letters*, vol. 31, no. 8, pp. 597–598, 1995.

[55] G. Balachandran and P. Allen, "Fully differential switched-current memory cell with low charge-injection errors," *IEE Proceedings-Circuits, Devices and Systems*, vol. 148, no. 3, pp. 157–164, 2001.

[56] Y. Sugimoto, "A realization of a below-1-V operational and 30-MS/s sample-and-hold IC with a 56-dB signal-to-noise ratio by applying the current-based circuit approach," *IEEE Transactions on Circuits and Systems I: Regular Papers*, vol. 51, no. 1, pp. 110–117, 2004.

[57] Y. Sugimoto and D. G. Haigh, "A current-mode circuit with a linearized input *V/I* conversion scheme and the realization of a 2-V/2.5-V operational, 100-MS/s, MOS SHA," *IEEE Transactions on Circuits and Systems I: Regular Papers*, vol. 55, no. 8, pp. 2178–2187, 2008.

[58] E. A. Vittoz, "Low-power design: Ways to approach the limits," in *Proc. 41$^{st}$ IEEE International Solid-State Circuits Conference, Digest of Technical Papers*, pp. 14–18, IEEE, 1994.

[59] C. D. Salthouse and R. Sarpeshkar, "A practical micropower programmable bandpass filter for use in bionic ears," *IEEE Journal of Solid-State Circuits*, vol. 38, no. 1, pp. 63–70, 2003.

[60] P. Corbishley and E. Rodriguez-Villegas, "A nanopower bandpass filter for detection of an acoustic signal in a wearable breathing detector," *IEEE Transactions on Biomedical Circuits and Systems*, vol. 1, no. 3, pp. 163–171, 2007.

[61] M. Tuckwell and C. Papavassiliou, "An analog gabor transform using sub-threshold 180-nm CMOS devices," *IEEE Transactions on Circuits and Systems I: Regular Papers*, vol. 56, no. 12, pp. 2597–2608, 2009.

[62] M. Yang, J. Liu, Y. Xiao, and H. Liao, "14.4 nW fourth-order bandpass filter for biomedical applications," *IET Electronics Letters*, vol. 46, no. 14, pp. 973–974, 2010.

[63] A. J. Casson and E. Rodriguez-Villegas, "A 60 pW gmC continuous wavelet transform circuit for portable EEG systems," *IEEE journal of solid-state circuits*, vol. 46, no. 6, pp. 1406–1415, 2011.

[64] R. Sarpeshkar, C. Salthouse, J.-J. Sit, M. W. Baker, S. M. Zhak, T.-T. Lu, L. Turicchia, and S. Balster, "An ultra-low-power programmable analog bionic ear processor," *IEEE Transactions on Biomedical Engineering*, vol. 52, no. 4, pp. 711–727, 2005.

[65] K. Suzuki, H. Miyashita, M. Suzuki, and C. Nakada, "Cochlear outer hair cell system is a logarithmic compressor," *Bioscience Hypotheses*, vol. 2, no. 2, pp. 69–74, 2009.

[66] D. Johns and K. Martin, *Analog integrated circuit design*. John Wiley & Sons. Inc., 1997.

[67] S. Solís-Bustos, J. Silva-Martínez, F. Maloberti, and E. Sánchez-Sinencio, "A 60-dB dynamic-range CMOS sixth-order 2.4-Hz low-pass filter for medical applications," *IEEE Transactions on Circuits and Systems II: Analog and Digital Signal Processing*, vol. 47, no. 12, pp. 1391–1398, 2000.

[68] A. Veeravalli, E. Sánchez-Sinencio, and J. Silva-Martínez, "Transconductance amplifier structures with very small transconductances: a comparative design approach," *IEEE Journal of Solid-State Circuits*, vol. 37, no. 6, pp. 770–775, 2002.

[69] X. Qian, Y. P. Xu, and X. Li, "A CMOS continuous-time low-pass notch filter for EEG systems," *Analog Integrated Circuits and Signal Processing*, vol. 44, no. 3, pp. 231–238, 2005.

[70] S.-Y. Lee and C.-J. Cheng, "Systematic design and modeling of a OTA-C filter for portable ecg detection," *IEEE Transactions on Biomedical Circuits and Systems*, vol. 3, no. 1, pp. 53–64, 2009.

[71] C. Mead, *Analog VLSI and neutral systems*, vol. 90. Reading, MA: Addison-Wesley Publishing Co., 1989.

[72] J. Webster, *Medical instrumentation: application and design*. John Wiley & Sons, 2009.

[73] J. García-Niebla and G. Serra-Autonell, "Effects of inadequate low-pass filter application," *Journal of Electrocardiology*, vol. 42, no. 4, pp. 303–304, 2009.

[74] P. Kligfield, L. S. Gettes, J. J. Bailey, *et al.*, "Recommendations for the standardization and interpretation of the electrocardiogram," *Journal of the American College of Cardiology*, vol. 49, no. 10, pp. 1109–1127, 2007.

[75] P. R. Rijnbeek, J. A. Kors, and M. Witsenburg, "Minimum bandwidth requirements for recording of pediatric electrocardiograms," *Circulation*, vol. 104, no. 25, pp. 3087–3090, 2001.

[76] L. Hejjel and L. Kellenyi, "The corner frequencies of the ECG amplifier for heart rate variability analysis," *Physiological measurement*, vol. 26, no. 1, p. 39, 2005.

[77] D. Python, A.-S. Porret, and C. Enz, "A 1 V 5th-order bessel filter dedicated to digital standard processes," in *Proc. IEEE Custom Integrated Circuits*, pp. 505–508, 1999.

[78] G. Groenewold, "Optimal dynamic range integrators," *IEEE Transactions on Circuits and Systems I: Fundamental Theory and Applications*, vol. 39, no. 8, pp. 614–627, 1992.

[79] Y. Tsividis, "Externally linear, time-invariant systems and their application to companding signal processors," *IEEE Transactions on Circuits and Systems II: Analog and Digital Signal Processing*, vol. 44, no. 2, pp. 65–85, 1997.

[80] E. A. Vittoz and Y. P. Tsividis, "Frequency-dynamic range-power," in *Trade-Offs in Analog Circuit Design*, pp. 283–313, Springer, 2002.

[81] Y. Li, C. C. Poon, and Y.-T. Zhang, "Analog integrated circuits design for processing physiological signals," *IEEE Reviews in Biomedical Engineering*, vol. 3, pp. 93–105, 2010.

[82] B. Linares-Barranco and T. Serrano-Gotarredona, "On the design and characterization of femtoampere current-mode circuits," *IEEE Journal of Solid-State Circuits*, vol. 38, no. 8, pp. 1353–1363, 2003.

[83] I. Pachnis, A. Demosthenous, and N. Donaldson, "Comparison of transconductance reduction techniques for the design of a very large time-constant CMOS integrator," in *Proc. 13th IEEE International Conference on Electronics, Circuits and Systems (ICECS'06)*, pp. 37–40, IEEE, 2006.

[84] M. Steyaert, P. Kinget, W. Sansen, and J. Van der Spiegel, "Full integration of extremely large time constants in CMOS," *Electronics Letters*, vol. 27, no. 10, pp. 790–791, 1991.

[85] P. Garde, "Transconductance cancellation for operational amplifiers," *IEEE Journal of Solid-State Circuits*, vol. 12, no. 3, pp. 310–311, 1977.

[86] A. El-Mourabit, G.-N. Lu, and P. Pittet, "Wide-linear-range subthreshold OTA for low-power, low-voltage, and low-frequency applications," *IEEE Transactions on Circuits and Systems I: Regular Papers*, vol. 52, no. 8, pp. 1481–1488, 2005.

[87] A. Tajalli and Y. Leblebici, "Low-power and widely tunable linearized biquadratic low-pass transconductor-C filter," *IEEE Transactions on Circuits and Systems II: Express Briefs*, vol. 58, no. 3, pp. 159–163, 2011.

[88] P. Bruschi, N. Nizza, F. Pieri, M. Schipani, and D. Cardisciani, "A fully integrated single-ended 1.5–15-Hz low-pass filter with linear tuning law," *IEEE Journal of Solid-State Circuits*, vol. 42, no. 7, pp. 1522–1528, 2007.

[89] E. Rodriguez-Villegas, A. J. Casson, and P. Corbishley, "A subhertz nanopower low-pass filter," *IEEE Transactions on Circuits and Systems II: Express Briefs*, vol. 58, no. 6, pp. 351–355, 2011.

[90] A. Becker-Gómez, U. Çilingiroglu, and J. Silva-Martinez, "Compact sub-hertz OTA-C filter design with interface-trap charge pump," *IEEE Journal of Solid-State Circuits*, vol. 38, no. 6, pp. 929–934, 2003.

[91] R. Rieger, A. Demosthenous, and J. Taylor, "A 230-nW 10-s time constant CMOS integrator for an adaptive nerve signal amplifier," *IEEE Journal of Solid-State Circuits*, vol. 39, no. 11, pp. 1968–1975, 2004.

[92] A. Arnaud, R. Fiorelli, and C. Galup-Montoro, "Nanowatt, sub-nS OTAs, with sub-10-mV input offset, using series-parallel current

mirrors," *IEEE Journal of Solid-State Circuits*, vol. 41, no. 9, pp. 2009–2018, 2006.

[93] R. R. Harrison, P. T. Watkins, R. J. Kier, R. O. Lovejoy, D. J. Black, B. Greger, and F. Solzbacher, "A low-power integrated circuit for a wireless 100-electrode neural recording system," *IEEE Journal of Solid-State Circuits*, vol. 42, no. 1, pp. 123–133, 2007.

[94] J. F. Kaiser, "On a simple algorithm to calculate the 'energy' of a signal," in *Proc. 1990 International Conference on Acoustics, Speech, and Signal Processing (ICASSP-90)*, pp. 381–384, IEEE, 1990.

[95] S. Mukhopadhyay and G. Ray, "A new interpretation of nonlinear energy operator and its efficacy in spike detection," *IEEE Transactions on Biomedical Engineering*, vol. 45, no. 2, pp. 180–187, 1998.

[96] J. Holleman, A. Mishra, C. Diorio, and B. Otis, "A micro-power neural spike detector and feature extractor in 0.13$\mu$m CMOS," in *Proc. IEEE Custom Integrated Circuits Conference (CICC 2008)*, pp. 333–336, IEEE, 2008.

[97] B. Gosselin and M. Sawan, "An ultra low-power cmos action potential detector," in *IEEE International Symposium on Circuits and Systems (ISCAS 2008)*, pp. 2733–2736, IEEE, 2008.

[98] B. Gosselin and M. Sawan, "Adaptive detection of action potentials using ultra low-power CMOS circuits," in *IEEE Biomedical Circuits and Systems Conference (BioCAS 2008)*, pp. 209–212, IEEE, 2008.

[99] C. Toumazou, F. J. Lidgey, and D. G. Haigh, eds., *Analog IC design: the current mode approach*. London, UK: Peregrinus, 1990.

[100] E. Yuce, "Design of a simple current-mode multiplier topology using a single CCCII+," *IEEE Transactions on Instrumentation and Measurement*, vol. 57, no. 3, pp. 631–637, 2008.

[101] C.-C. Chang and S.-I. Liu, "Weak inversion four-quadrant multiplier and two-quadrant divider," *IEE Electronics Letters*, vol. 34, no. 22, pp. 2079–2080, 1998.

[102] A. G. Katsiamis, K. N. Glaros, and E. M. Drakakis, "Insights and advances on the design of CMOS sinh companding filters," *IEEE Transaction on Circuits and Systems-I: Regular Papers*, vol. 55, no. 9, pp. 2539–2550, 2008.

[103] T. Serrano-Gotarredona, B. Linares-Barranco, and A. G. Andreou, "A general translinear principle for subthreshold MOS transistors," *IEEE Transactions on Circuits and Systems I: Fundamental Theory and Applications*, vol. 46, no. 5, pp. 607–616, 1999.

[104] B. S. Wilson, C. C. Finley, D. T. Lawson, R. D. Wolford, D. K. Eddington, and W. M. Rabinowitz, "Better speech recognition with cochlear implants," *Nature*, vol. 352, no. 6332, pp. 236–238, 1991.

[105] Y.-Y. Kong, R. Cruz, J. A. Jones, and F.-G. Zeng, "Music perception with temporal cues in acoustic and electric hearing," *Ear and Hearing*, vol. 25, no. 2, pp. 173–185, 2004.

[106] I. Tasaki, "Nerve impulses in individual auditory nerve fibers of guinea pig," *J. Neurophysiol*, vol. 17, no. 97, p. 122, 1954.

[107] N. Y.-S. Kiang, "Discharge patterns of single fibers in the cat's auditory nerve.," tech. rep., DTIC Document, 1965.

[108] F.-G. Zeng, "Trends in cochlear implants," *Trends in amplification*, vol. 8, no. 1, pp. 1–34, 2004.

[109] S. Padala and K. Prabhu, "Systolic arrays for the discrete hilbert transform," in *IEE Proceedings on Circuits, Devices and Systems*, vol. 144, pp. 259–264, IET, 1997.

[110] W. Ngamkham, C. Sawigun, S. Hiseni, and W. A. Serdijn, "Analog complex gammatone filter for cochlear implant channels," in *Proceedings of 2010 IEEE International Symposium on Circuits and Systems (ISCAS)*, pp. 969–972, IEEE, 2010.

[111] J.-J. Sit, A. M. Simonson, A. J. Oxenham, M. A. Faltys, and R. Sarpeshkar, "A low-power asynchronous interleaved sampling algorithm for cochlear implants that encodes envelope and phase information," *IEEE Transactions on Biomedical Engineering*, vol. 54, no. 1, pp. 138–149, 2007.

[112] P. Schleich, "Pulsatile cochlear implant stimulation strategy," Oct. 22 2009. WO Patent 2,009,062,142.

[113] K. Nie, A. Barco, and F.-G. Zeng, "Spectral and temporal cues in cochlear implant speech perception," *Ear and hearing*, vol. 27, no. 2, pp. 208–217, 2006.

[114] J. Chen, X. Wu, L. Li, and H. Chi, "Simulated phase-locking stimulation: An improved speech processing strategy for cochlear implants," *ORL*, vol. 71, no. 4, pp. 221–227, 2009.

[115] B. S. Wilson, D. T. Lawson, M. Zerbi, and C. Finley, "Speech processors for auditory prostheses," *NIH project*, no. 01, 1994.

[116] M. W. White, M. M. Merzenich, and J. N. Gardi, "Multichannel cochlear implants: Channel interactions and processor design," *Archives of Otolaryngology*, vol. 110, no. 8, p. 493, 1984.

[117] B. R. Glasberg and B. C. Moore, "Derivation of auditory filter shapes from notched-noise data," *Hearing research*, vol. 47, no. 1, pp. 103–138, 1990.

[118] R. Wessel, C. Koch, and F. Gabbiani, "Coding of time-varying electric field amplitude modulations in a wave-type electric fish," *Journal of neurophysiology*, vol. 75, no. 6, pp. 2280–2293, 1996.

[119] K. Koli and K. Halonen, "Low voltage MOS-transistor-only precision current peak detector with signal independent discharge time constant," in *Proceedings of 1997 IEEE International Symposium on Circuits and Systems (ISCAS'97)*, vol. 3, pp. 1992–1995, IEEE, 1997.

[120] R. Dlugosz and K. Iniewski, "High-precision analogue peak detector for X-ray imaging applications," *IET Electronics Letters*, vol. 43, no. 8, pp. 440–441, 2007.

# Index

# About the Authors

**Chutham Sawigun** was born in Udonthani, Thailand in 1978. He received B.Eng. (Electrical Power Engineering), M.Eng. (Electronic Engineering) and PhD (Analog IC Design for Biomedical Applications) from Ubonratchathanee University, Mahanakorn University of Technology and Delft University of Technology, respectively.

Currently, he is conducting research in the area of nanopower IC design for biomedical recording and stimulation in the Centre for Bioelectronics Integrated Systems, Mahanakorn University of Technology, Thailand. At the same time, he is an assistant professor in the Department of Electronic Engineering at the same university.

He also serves as a reviewer for *IEEE Transaction on Circuits and Systems I&II, IEEE Journal of Solid-State Circuits, IET Circuits, Devices and Systems, Microelectronic Journal* and *Analog Integrated Circuits and Signal Processing International Journal.* He is a conference committee for the International Conference on Biomedical Electronics and Devices: *BIODEVICES/BIOSTEC* and International Congress of Cardiovascular Technologies: *CARDIOTECHNIX.* In 2005 and 2012, he received the best paper award from The 12th International Conference on Electrical Engineering/Electronics, Telecommunications and Information Technology *(ECTICON-2005)* and the 35th Electrical Engineering Conference *(EECON-35),* respectively.

**Wouter A. Serdijn** (M'98, SM'08, F'11) was born in Zoetermeer ('Sweet Lake City'), the Netherlands, in 1966. He received the M.Sc. (cum laude) and Ph.D. degrees from Delft University of Technology, Delft, The Netherlands, in 1989 and 1994, respectively. Currently, he is a full professor in bioelectronics at Delft University of Technology, where he heads the Section Bioelectronics.

His research interests include integrated biomedical circuits and systems for biosignal conditioning and detection, neuroprosthetics, transcutaneous wireless communication, power management and energy harvesting as applied in, e.g., hearing instruments, cardiac pacemakers, cochlear implants, neurostimulators, portable, wearable, implantable and injectable medical devices and electroceuticals.

He is co-editor and co-author of the books Design of Efficient and Safe Neural Stimulators – a multidisciplinary approach (Springer, 2016), EMI-Resilient Amplifier Circuits (Springer 2013), Ultra Low-Power Biomedical Signal Processing: an analog wavelet filter approach for pacemakers (Springer, 2009), Circuits and Systems for Future Generations of Wireless Communications (Springer, 2009), Power Aware Architecting for data dominated applications (Springer, 2007), Adaptive Low-Power Circuits for Wireless Communications (Springer, 2006), Research Perspectives on Dynamic Translinear and Log-Domain Circuits (Kluwer, 2000), Dynamic Translinear and Log-Domain Circuits (Kluwer, 1998) and Low-Voltage Low-Power Analog Integrated Circuits (Kluwer, 1995). He authored and co-authored 8 book chapters, 2 patents and more than 300 scientific publications and presentations. He teaches Circuit Theory, Analog Integrated Circuit Design, Analog CMOS Filter Design, Active Implantable Biomedical Microsystems and Bioelectronics. He received the Electrical Engineering Best Teacher Award in 2001, in 2004 and in 2015.

He has served, a.o., as General Co-Chair for IEEE ISCAS 2015 and for IEEE BioCAS 2013, Technical Program Chair for IEEE BioCAS 2010 and for IEEE ISCAS 2010, 2012 and 2014, as a member of the Board of Governors

(BoG) of the IEEE Circuits and Systems Society (2006–2011), as chair of the Analog Signal Processing Technical Committee of the IEEE Circuits and Systems society, as a member of the Steering Committee of the IEEE Transactions on Biomedical Circuits and Systems (T-BioCAS) and as Editor-in-Chief for IEEE Transactions on Circuits and Systems—I: Regular Papers (2010–2011).

Wouter A. Serdijn is an IEEE Fellow, an IEEE Distinguished Lecturer and a mentor of the IEEE.

Lightning Source UK Ltd.
Milton Keynes UK
UKOW06n1146270117

293027UK00002B/14/P